Springer Theses

Recognizing Outstanding Ph.D. Research

For further volumes:
http://www.springer.com/series/8790

Aims and Scope

The series "Springer Theses" brings together a selection of the very best Ph.D. theses from around the world and across the physical sciences. Nominated and endorsed by two recognized specialists, each published volume has been selected for its scientific excellence and the high impact of its contents for the pertinent field of research. For greater accessibility to non-specialists, the published versions include an extended introduction, as well as a foreword by the student's supervisor explaining the special relevance of the work for the field. As a whole, the series will provide a valuable resource both for newcomers to the research fields described, and for other scientists seeking detailed background information on special questions. Finally, it provides an accredited documentation of the valuable contributions made by today's younger generation of scientists.

Theses are accepted into the series by invited nomination only and must fulfill all of the following criteria

- They must be written in good English.
- The topic of should fall within the confines of Chemistry, Physics and related interdisciplinary fields such as Materials, Nanoscience, Chemical Engineering, Complex Systems and Biophysics.
- The work reported in the thesis must represent a significant scientific advance.
- If the thesis includes previously published material, permission to reproduce this must be gained from the respective copyright holder.
- They must have been examined and passed during the 12 months prior to nomination.
- Each thesis should include a foreword by the supervisor outlining the significance of its content.
- The theses should have a clearly defined structure including an introduction accessible to scientists not expert in that particular field.

Matthias Heydt

How Do Spores Select Where to Settle?

A Holographic Motility Analysis of *Ulva* Zoospores on Different Surfaces

Doctoral Thesis accepted by
University of Heidelberg, Germany

 Springer

Author
Dr. Matthias Heydt
Department of Applied Physical Chemistry
University of Heidelberg
Im Neuenheimer Feld 253
69120 Heidelberg
Germany
e-mail: m.heydt@web.de

Supervisor
Prof. Dr. Michael Grunze
Department of Applied Physical Chemistry
University of Heidelberg
Im Neuenheimer Feld 253
69120 Heidelberg
Germany
e-mail: michael.grunze@urz.uni-heidelberg.de

ISSN 2190-5053

e-ISSN 2190-5061

ISBN 978-3-642-17216-8

e-ISBN 978-3-642-17217-5

DOI 10.1007/978-3-642-17217-5

Springer Heidelberg Dordrecht London New York

Cover design: eStudio Calamar, Berlin/Figueres

Printed on acid-free paper

Springer is part of Springer Science+Business Media (www.springer.com)

For Anouk

Parts of This Thesis Have been Published in the Following Journal Articles

Holographic view on biofouling—How algal spores select a surface for settlement, M. Heydt, M. Pettitt, X. Cao, M.E. Callow, J.A. Callow, M. Grunze, A. Rosenhahn, 2010 (submitted).

Flow conditions in the vicinity of microstructures interfaces studied by holography and implications for assembly of artificial actin networks, S. Weiße, M. Heydt, T. Meier, S. Schulz, J.P. Spatz, M. Grunze, T. Haraszti, A. Rosenhahn. 2010 (submitted).

Classification of swimming microorganisms movements in 4D digital in-line holography data, L. Leal Taixé, M. Heydt, S. Weiße, A. Rosenhahn, B. Rosenhahn. DAGM 32nd Annual Symposium of the German Association for Pattern Recognition, 2010, Darmstadt, Germany.

Automatic tracking of swimming microorganisms in 4D digital in-line holography data, L. Leal Taixé, M. Heydt, A. Rosenhahn, B. Rosenhahn. IEEE Workshop on Motion and Video Computing (WMVC), 2009, Snowbird, Utah, USA.

Analysis of holographic microscopy data to quantitatively investigate three-dimensional settlement dynamics of algal zoospores in the vicinity of surfaces, M. Heydt, P. Divós, M. Grunze, A. Rosenhahn. The European Physical Journal E, 2009, 30: 141–148.

Digital in-line holography as a three-dimensional tool to study motile marine organisms during their exploration of surfaces, M. Heydt, A. Rosenhahn, M. Grunze, M. Pettitt, M.E. Callow, J.A. Callow. Journal of Adhesion, 2007, **83**(5): 417–430.

Supervisor's Foreword

Biofouling, the colonization of submerged natural or man-made surfaces by marine organisms, causes major problems for various industries. Aquaculture systems, heat exchangers, under water sensors, membrane filters, and ships are only some examples prone to biofouling. Many different organisms are involved in this process and interact during colonization. One particular biofouling model organism is the quadriflagellated zoospore of *Ulva linza*. Due to its motility the spores explore surfaces and are highly discriminative in selecting a surface for settlement to subsequently grow into macroscopic visible plants. For a more detailed understanding of the initial surface exploration phase, which leads to the irreversible growth on the surface, Matthias Heydt studied the exploration and settlement behavior by digital in-line holography in real time and three dimensional (3D). For his work "How do spores select where to settle? A holographic motility analysis of *Ulva* zoospores on different surfaces" he developed a transportable digital in-line holographic microscope and recorded the motility of freshly harvested *Ulva* spores during the settlement phase. Custom software packages were developed to reconstruct the holograms and to allow an automated determination of spore positions. Using the obtained trajectories, the motility of *Ulva* spores in solution and in the vicinity of surfaces is for the first time quantitatively analyzed in 3D. Functionalized glass surfaces with different wettability and different attractiveness for spore settlement are used as test substrates: Poly(ethylene glycol) (PEG), hydrophilic glass (AWG) and hydrophobic glass functionalized with fluorooctyltriethoxysilane (FOTS). Spores accumulate in a 200 µm wide surface boundary layer. On these different surface chemistries, a significantly different behavior is found including different motion patterns and also different velocity and swimming angle distributions. Analysis of the data recorded in the 2 min after the start of the tracking experiment allows prediction of the outcome of much longer classical laboratory assays. Especially the obtained results on the PEG surface clearly show that spores experience a repulsive interaction in the proximity of the surface. On fluorinated surfaces a time dependence of settlement attempts and patterns are observed, which are connected to the conditioning of hydrophobic surfaces by organic molecules and matter from Solution.

Holography is a new approach to understand biofouling and the presented thesis demonstrate its usefulness as it allows to quantitatively describe the interaction between organisms and surfaces leading to biofilm formation. This approach provides a mechanistic understanding and will help to develop new approaches for environmental benign coatings based on rational design. The obtained results inspired the work on a fully automated trajectory determination software, which will in future give access to a large motility data base. Meanwhile the work has been extended to study the swarming behavior of bacteria and host-pathogen interactions of Trypanosoma. These applications are only the start as understanding of 3D motility is of great importance for surface colonization also by other species, developmental biology, host-pathogen interactions, and predator-prey interactions. We expect that the unique capability of in-line holography to quantitatively analyze and thus understand motility will find many users in the future.

Heidelberg, December 2010   Axel Rosenhahn
                                                                 Michael Grunze

Acknowledgements

thanks to...

- Prof. Dr. Michael Grunze for the possibility to join his group, the confidence in me and my work, the countless ideas, the good mentoring, the invitations to conferences and the freedom to do my work independently.
- Prof. Dr. James Callow and Dr. Maureen Callow for the great possibility to be a participant of the AMBIO project. Furthermore, many thanks for useful discussions and your hospitality during my research stays at the University of Birmingham, England, UK.
- Dr. Axel Rosenhahn for the patient and extensive mentoring, the countless talks and discussions, the great time and for my first surf lessons in Rio.
- Dr. Michala Pettitt for the help with the organization and the hospitality during my four trips to the University of Birmingham, England. Many thanks for the great explanation of the secrets of how to work with *Ulva* spores. I am grateful for all the discussions about the thesis and your ironic attitude to cope with silly questions of a chemist trying to understand biology.
- Sebastian Weiße for the extensive checking and debugging of the analysis software ("Matthias, I broke it again"). Thanks for the numerous discussions about the relevance of the results and for the spell check which surely must have driven you crazy from time to time.
- Ruth Heine who helped with her critical and technical suggestions to finally finish this thesis. Also thanks for your friendship and the relaxing coffee breaks.
- Peter Divós who introduced me into programming in Matlab.
- Xinyu Cao who prepared all chemical modified surfaces. Even at the time when he was supposed to finish his thesis he prepared some additional PEG surfaces while I was already in Birmingham. Afterwards it turned out that these samples contributed to the major results of my thesis. Thank you very much; I will never forget this favor.
- Celine Rüdiger who helped me with the boring trajectory determination.
- Prof. Dr. Hans-Jürgen Kreuzer for the help with acquiring my first holograms and the hospitality during my research stay in Canada.

- Prof. Dr. Manfred Jericho and Stefan Jericho for the hospitality and all the help during my research stays at Dalhousie University, Halifax, Canada.
- Sören Schilp for the great time together: on conferences, on the 20 h journeys in the "Ulva-Mobil" with and without radio, on the four times 4 weeks research stays in Birmingham or on the shared surf adventure with a miracle healer at the pizza snack bar in Florida, USA.
- all, who contributed in any way to the proof reading of the thesis:
  Mike Beckers, Ruth Heine, Henry Heydt, Katja Hoffmann, Christina Leinweber, Michala Pettitt, Axel Rosenhahn, Sören Schilp, Martina Schürmann, Svenja Stuppy, Isabel Thome und Sebastian Weiße.
- Dr. André Khalil for his nice lecture and his great idea to enhance the precision of the z-position determination.
- Florian Staier for the help with LabView.
- Reinhold Jehle for his assistance in CAD drawings.
- the workshop team at the institute for the production of the needed parts—often on short notice.
- Peter Jeschka and Günther Meinusch for the advices and the assistance in the area of electronics and data processing.
- the team of the administration: Swetlana Duchnay, Karin Jordan, Andrea Miech und Benjamin Scherke for the contribution to the many important but also annoying things.
- all colleagues in the group for the nice working atmosphere.
- the old crew of INF 229, especially to Martina Schürmann und Ruth Heine, for a good start in the PhD.
- my friends, who reminded me that there is something else in the world than sitting in front of a computer and clicking on white dots.
- my parents who have always supported me and on whom I can always rely on.
- Katja who had to cope with much frustration during the last year. Nevertheless she was always a great support and an endless source of inspiration. Thank you for the great time together.

Contents

List of Abbreviations

$\bar{\alpha}_v$	mean angle between two consecutive vectors
$\bar{\alpha}_z$	mean angle towards the surface normal
af	angular frequency (β/dt)
AMBIO	Advanced Nanostructured Surfaces for the Control of Biofouling
app.	approach to the surface
ASM	active searching motion: spore fraction assigned to the *gyration, hit and run, orientation* pattern and detachment and approach for spores assigned to the *hit and stick* pattern
ASW	artificial sea water
AWG	acid washed glass
CW	clockwise
CCW	counter clockwise
det.	detach from the surface
DIH	digital in-line holography
DIHM	digital in-line holography microscopy
dist.	distance away from the surface
E. coli	*Escherichia coli*
EG_6	hexa(ethylene glycol)-containing SAMs
EPS	adhesive vesicles
FOTS	per-fluorinated coating on glass
FoV	field of view
FWHM	full width at half-maximum
GB	giga byte
Gy	gyration
h	hour
H&R	hit and run
H&S	hit and stick
HWHM	half width at half-maximum
L11	lower reconstruction distance
L12	higher reconstruction distance
min	minute

NA	numerical aperture
Ox-PDMS	oxidized PDMS
PDMS	polydimethylsiloxan
PEG	PEG2000 coating on glass
ra	radius [μm]
Re	Reynolds number
rf	rotation frequency
RSF	ratio between the amount of slow and fast spores
s	second
SEM	scanning electron microscopy
Sp	spinning
v_m	mean velocity
v_p	most probable velocity
η	dynamic viscosity [Nsec/m^2]
α_v	angle between two consecutive vectors
α_z	angle towards the surface normal
λ	wave length
σ	standard deviation

Chapter 1
Introduction

Marine biofouling can be defined as the undesirable growth of microorganisms, plants and animals on submerged surfaces. Biofouling causes severe economics cost, e.g. because of the increased drag of a moving vessel which results in a higher fuel consumption. For the US Navy alone the annual extra cost caused by fouling on ship hulls is estimated to be 1 billion US$ [1].

Over the decades scientists and engineers have tried to find an effective way to prevent fouling on man-made structures. While in the past success has been achieved with toxic surfaces (e.g. tributyltin (TBT)-based paint) by killing the settling organisms, nowadays these surfaces are restricted by the European Union (E.U.) and International Maritime Organization (IMO) legislation because of the large environmental side effect on the marine ecosystem and accumulation of the toxic ingredients in the food chain [2, 3]. Finding an environmentally friendly antifouling surface has been a major field of research over the last years. Even though many design guidelines for the latter have been published recently it still remains difficult to develop an effective antifouling coating [4–11]. The reason for that is on the one hand the very diverse range of fouling marine organisms (e.g. bacteria, algae, barnacles) with stages of settling spanning several orders of magnitude, ranging from hundreds of nanometers to centimeters. On the other hand, to design an antifouling coating endeavors knowledge from the field of surface science, biomaterial science, marine biology, organic chemistry and engineering. The aim of the EC Framework 6 Integrated Project AMBIO "(Advanced Nanostructured Surfaces for the Control of Biofouling)" [12] is to find new approaches to design environmentally friendly antifouling coatings and therefore combines industry and university knowledge throughout the above mentioned fields. The here presented work is funded by this project.

One environmentally friendly possibility to keep a ship hull fairly clean is the use of fouling-release coatings (e.g. Intersleek® International Paint) which can be cleaned by applying a weak mechanical force, such as the one supplied by shear forces acting on a boat in motion [13]. Nevertheless these types of coatings can only be applied on ships which do not stay in a harbor for longer periods. Once a

M. Heydt, *How Do Spores Select Where to Settle?*, Springer Theses,
DOI: 10.1007/978-3-642-17217-5_1, © Springer-Verlag Berlin Heidelberg 2011

certain mass of fouling has accumulated on the coating it is hard to remove and the coating has lost its fouling release properties.

The result of biofouling is easily recognizable on a macroscopic level, but the initial process which leads to fouling occurs on a small length scale which depends on the size of the fouling organism [14]. The length scale for bacteria for example is hundreds of nanometer, whereas for the algae *Ulva*, fouling occurs on a length scale of 5 μm. The experiments in the course of this thesis are exclusively done with spores of the algae *Ulva linza* commonly known as seaweed. *Ulva linza* is used in AMBIO as a model organism to represent soft-macrofouling species.

The performance of a potential antifouling surface is rated by two factors. First, the amount of settlement is determined after a defined time span and second, the adhesion strength of the attached organisms is analyzed. For microscopic organisms (bacteria and algae) this strength is typically determined by shear stress [15], for larger organisms (mussels, barnacles) a manual or automated force gauge [16] are used.

Even though the performance of an antifouling coating can be described based on these two factors, the dynamics in the exploration behavior are not accessible using this approach. For organisms in the size range of hundreds of micrometers (e.g. barnacles) it is possible to study the exploration behavior by video microscopy [17–19]. However, for smaller organisms (e.g. bacteria and algae) tracking by video microscopy gets more challenging. For example in the work of Berg it was only possible to follow one microorganism at a time [20, 21]. Iken et al. studied the swimming behavior of spores of brown algae in a 10 μl drop between a microscope slide and 22 × 44 mm cover slip which was supported at both ends by a cover slip to allow spore movement [22, 23]. With this setup they reduced the natural three dimensional (3D) motion into a two dimensional (2D) movement and could therefore acquire trajectories of many organisms simultaneously. Nevertheless, they observed several different swimming patterns and were able to correlate these to different spore behavior.

In the course of this thesis, digital in-line holographic microscopy (DIHM) is used to obtain not only 2D but the real 3D movement of spores. DIHM goes back to the initial idea of holography proposed by Gabor in 1940s [24]. In the last decades Kreuzer successfully adapted this idea for LASER radiation with digital data acquisition and implemented a fast and accurate numerical reconstruction algorithm [25, 26]. Furthermore, he could demonstrate the applicability of this technique to study marine organisms as algae [27, 28] and developed an instrument for in situ measurements in the ocean [29].

The aim of this work is to study the 3D exploration behavior of algae spores near surfaces using DIHM and correlate this behavior to the known antifouling performance of the surfaces. It is known that *Ulva* spores can select their surface position prior settlement. Whether the exploration behavior is influenced by the surface properties is studied in this thesis. The major effort during the course of the thesis was to develop semi-automatic software to determine the motion trajectories from the measured data. The software was programmed in MATLAB® and was used to process large amounts of motion data. Within the course of this thesis the motility of *Ulva* spores is characterized in detail in 3D for the first time. The motility is analyzed in solution and in vicinity to different surfaces.

In the following chapter (Chap. 2), an introduction to the basic theory of holography is provided. To put the work of this thesis in a general context the existing literature is reviewed in the next section (Chap. 3). Since the focus of the work is on the analysis of swimming behavior, the developed software is described together with the experimental details in Chap. 4. The results obtained for the motility are presented in the next chapter (Chap. 5). The chapter is organized the way that first the motility in solution is explained and discussed before the behavior in vicinity to different surfaces is concisely presented. The full and detailed analysis of the surface exploration behavior is provided in the appendix (9.1 AWG, 9.2 PEG, and 9.3 FOTS). The observed surface exploration behavior is discussed in an independent Chap. 6. Finally, an outlook for future experiments and experimental development concludes the thesis.

References

1. M.E. Callow, J.A. Callow, Biologist **49**(1), 1–5 (2002)
2. D.M. Yebra, S. Kiil, K. Dam-Johansen, Prog. Org. Coat. **50**(2), 75–104 (2004)
3. B. Antizar-Ladislao, Environ. Int. **34**(2), 292–308 (2008)
4. H. Monfared, F. Sharif, Prog. Org. Coat. **63**(1), 79–86 (2008)
5. A.J. Scardino, D. Hudleston, Z. Peng, N.A. Paul, R. de Nys, Biofouling **25**(1), 83–93 (2009)
6. J. Genzer, K. Efimenko, Biofouling **22**(5), 339–360 (2006)
7. C.M. Grozea, G.C. Walker, Soft Matter **5**(21), 4088–4100 (2009)
8. N. Aldred, A.S. Clare, Biofouling **24**(5), 351–363 (2008)
9. L.D. Chambers, K.R. Stokes, F.C. Walsh, R.J.K. Wood, Surf. Coat. Technol. **201**(6), 3642–3652 (2006)
10. A. Whelan, F. Regan, J. Environ. Monitor. **8**(9), 880–886 (2006)
11. S. Dobretsov, H.U. Dahms, P.Y. Qian, Biofouling **22**(1), 43–54 (2006)
12. EC Framework 6 Integrated Project "AMBIO"—Advanced Nanostructured Surfaces for the Control of Biofouling. http://www.ambio.bham.ac.uk. Accessed 23 Nov 2009
13. C.J. Kavanagh, R.D. Quinn, G.W. Swain, J. Adhesion **81**(7–8), 843–868 (2005)
14. A. Rosenhahn, T. Ederth, M.E. Pettitt, Biointerphases **3**(1), IR1–IR5 (2008)
15. M.P. Schultz, J.A. Finlay, M.E. Callow, J.A. Callow, Biofouling **19**, 17–26 (2003)
16. S.L. Conlan, R.J. Mutton, N. Aldred, A.S. Clare, Biofouling **24**(6), 471–481 (2008)
17. G.S. Prendergast, C.M. Zurn, A.V. Bers, R.M. Head, L.J. Hansson, J.C. Thomason, Biofouling **24**(6), 449–459 (2008)
18. J.P. Marechal, C. Hellio, M. Sebire, A.S. Clare, Biofouling **20**(4–5), 211–217 (2004)
19. M.O. Amsler, C.D. Amsler, D. Rittschof, M.A. Becerro, J.B. McClintock, Marine Freshw. Behav. Physiol. **39**(4), 259–268 (2006)
20. H.C. Berg, *Random Walks in Biology* (Princeton University Press, Princeton, 1993)
21. H.C. Berg, Rev. Sci. Instrum. **42**(6), 869–871 (1971)
22. K. Iken, C.D. Amsler, S.R. Greer, J.B. McClintock, Phycologia **40**(4), 359–366 (2001)
23. K. Iken, S.P. Greer, C.D. Amsler, J.B. McClintock, Biofouling **19**(5), 327–334 (2003)
24. D. Gabor, Nature **161**(8), 777 (1948)
25. H.J. Kreuzer, M.H. Jericho, W. Xu, Proc. SPIE **4401**, 234–244 (2001)
26. H. J. Kreuzer, U.S. Patent, 6.411.406, 2002

27. W. Xu, M.H. Jericho, H.J. Kreuzer, I.A. Meinertzhagen, Optics Lett. **28**(3), 164–166 (2003)
28. N.I. Lewis, W.B. Xu, S.K. Jericho, H.J. Kreuzer, M.H. Jericho, A.D. Cembella, Phycologia **45**(1), 61–70 (2006)
29. S.K. Jericho, J. Garcia-Sucerquia, W.B. Xu, M.H. Jericho, H.J. Kreuzer, Rev. Scientific Instrum. **77**(4) 8 pp (2006)

Chapter 2
Theory of Holography

Holography has been used as a tool to determine the 3D motion data of swimming microorganisms hence the basic principle of holography is briefly explained in this chapter. A more detailed explanation of holography can be found in the Ph.D. theses of Dr. R. Barth [1] and Dr. M. Schürmann [2] and in the Diploma thesis of T. Gorniak [3] all carried out in our group. Furthermore, this introduction is based on the corresponding chapters in common textbooks [4–10].

To understand the basic holography principles some general wave phenomena are briefly explained in the following.

2.1 Properties of Light Waves

2.1.1 Intensity

When detecting a light wave the crucial parameter is the intensity of the wave. This is true for a human eye as well as for other detectors [11]. A light wave can be described by the wavefunction [5]

$$\psi(\vec{r}) = A(\vec{r}) \, \exp\{i\phi(\vec{r})\} \tag{2.1}$$

where $A(\vec{r})$ is the amplitude and $\phi(\vec{r})$ is the phase. The intensity is defined as the square of the absolute value of the wave function

$$I(\vec{r}) = |\psi(\vec{r})|^2 \tag{2.2}$$

In general, the intensity describes how much energy per time is transported to a plane perpendicular to the wave vector. For plane (I_p) and spherical ($I_s(\vec{r})$) waves it is [10, 11]

$$I_p = |A|^2 \text{ and } I_s(\vec{r}) = \frac{1}{r^2}|A|^2 \tag{2.3}$$

M. Heydt, *How Do Spores Select Where to Settle?*, Springer Theses,
DOI: 10.1007/978-3-642-17217-5_2, © Springer-Verlag Berlin Heidelberg 2011

2.1.2 Interference

Interference is the superposition of two or more waves. Since the wave equation is a linear differential equation the resulting wave function is the linear combination of the individual functions [5]. For two monochromatic waves $(\psi_1, \psi_2(\vec{r}))$ with equal frequency and polarization the total intensity is

$$I(\vec{r}) = |\psi_1(\vec{r}) + \psi_2(\vec{r})|^2 = I_1 + I_2 + 2\sqrt{I_1 + I_2}\cos\{\phi\}. \qquad (2.4)$$

The individual intensities are I_1 and I_2 and the phase difference is $\phi = \phi_1 - \phi_2$. Apparently the total intensity is not the simple sum of the individual intensities, but the so called interference term $2\sqrt{I_1 I_2}\cos\{\phi\}$ has to be added. This term can be positive (constructive interference) or negative (destructive interference) and causes the modulation of the intensity visible as dark and bright fringes [5, 10].

2.1.3 Coherence

The basis for interference phenomena is a constant correlation of the phases of the individual waves. This correlation is called coherence and it is distinguished between temporal and spatial coherence. Temporal coherence is understood as the measure of the average correlation for a wave at two points in time (separated by a delay). If the source is not point like but rather extended in one or two dimensions, spatial coherence describes the ability of two points in space to interfere with each other [5].

2.1.4 Diffraction

Another important phenomenon for (light) waves apart from interference is diffraction. If a wave encounters an obstacle with dimensions in the range of its wavelength diffraction occurs which cannot be explained by geometrical optics. The Huygens' Principle gives a qualitative explanation for diffraction. It states that "every point of a wave front can be considered as a point source for a secondary wave. The wave front at any other place is the coherent superposition of these secondary waves" [5]. When a circular aperture (e.g. a 500 nm pinhole) is used with visible light a spherical wave is generated by diffraction which can be used for illumination in a holographic experiment.

2.2 Holography

Standard photography is widely used to conserve moments, but it has the disadvantage that only a two-dimensional projection of the three-dimensional world is stored. Conventional recording media (e.g. CCD-chip, photo plate, …) only respond to the intensity of the light waves. Therefore, according to Eq. (2.12), the phase information ϕ is lost in the image storing process. If the amplitude A as well as the phase ϕ of a wave front in an image can be reproduced, a perfect image of the object is generated which is impossible to distinguish from the original.

2.2.1 Principle

Denis Gabor observed a possibility to record the phase additionally to the amplitude and subsequently to reconstruct the object wave [12]. Since any light sensitive media—as described above—is only able to store the amplitude of a wave, Gabor encoded the phase information of the object wave ψ_{obj} by recording an interference pattern of the object wave ψ_{obj} and a reference wave ψ_{ref}. According to Eq. (2.4) the observed intensity of the interference pattern is

$$
\begin{aligned}
I(x,y) &= \left|\psi_{ref} + \psi_{obj}\right|^2 \\
&= \psi_{ref}^*\psi_{ref} + \psi_{ref}^*\psi_{obj} + \psi_{ref}\psi_{obj}^* + \psi_{obj}^*\psi_{obj} \\
&= A_{ref}^2 + A_{obj}^2 + 2A_{ref}A_{obj}\,\cos\{\phi_{ref} - \phi_{obj}\} \\
&= I_{ref} + I_{obj} + 2\sqrt{I_{ref}I_{obj}}\,\cos\{\phi_{ref} - \phi_{obj}\}
\end{aligned}
\tag{2.5}
$$

and is called a hologram [5, 10]. The recorded object can be reconstructed either by illumination of the reference wave [4, 5, 10] ψ_{ref} or by a numerical reconstruction [13]. The latter is widely used today and makes holography feasible for many applications [14–22].

Holography can be performed by using plane or spherical reference waves and it is also possible to work in different setup geometries. Gabor himself used the so called in-line geometry. This setup is also used in the course of this thesis and is therefore introduced in the following. The other geometries like off-axis and Fourier geometries are not discussed but a detailed description can be found in literature [1, 4, 5].

2.2.2 In-line Holography

The characteristics of the in-line geometry are that the source of the reference, the sample and the recording screen are placed on one axis. Typically, the recording screen is orientated perpendicular to the optical axis. In Fig. 2.1, a schematic drawing of Gabors In-line setup is shown.

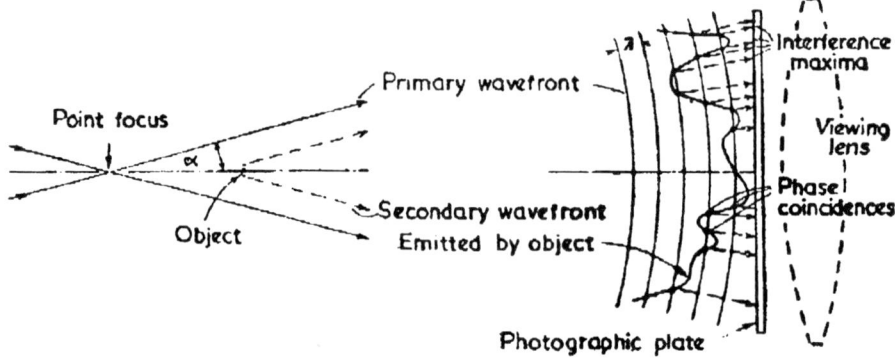

Fig. 2.1 In-line holography setup as published by Gabor in his article "A New Microscope Principle" [12]

The object of interest is placed in a certain distance to the recording screen and is illuminated with a spherical reference wave ψ_{ref} propagating from the source towards the screen. The wave which is diffracted at the sample forms the object wave ψ_{obj} which interferes with the undiffracted part of the reference wave ψ_{ref}. The resulting interference pattern is observed on the recording screen and is called a hologram.

The easiest way to achieve a magnification in the in-line geometry is to use a spherical reference wave for recording and a plane wave with the same wavelength during the reconstruction. If the distance of the recording screen to the point source is L (see Fig. 2.2), and the object is placed in a distance l to the point source, the magnification is [5, 7]

$$M = \frac{L}{l}. \tag{2.6}$$

Fig. 2.2 Schematic drawing of an in-line geometry setup. The pinhole detector distance is L. The sample (*gray*) is positioned in a distance l from the pinhole and the detector size is D. The reference wave φ_{ref} is illustrated in *black*, dashed and the object wave φ_{obj} in *gray*, dotted. The angle\alpha denotes the half-angle of beam spread

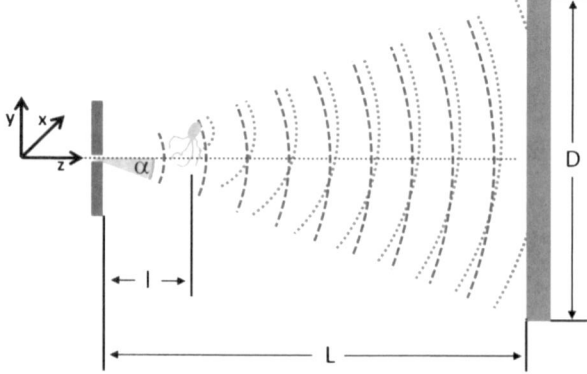

Using this approach and especially in combination with a numerical recon-struction, in-line holography can be used as a powerful microscopy technique [15, 17, 18, 20].

2.2.3 Numerical Reconstruction

In digital holography the hologram is recorded digitally and can be reconstructed numerically. To perform this calculation the propagation and thus the complex amplitude of the wave is computed numerically. The derivation of the wave front in a specific plane is called numerical reconstruction [13]. These reconstruction planes correspond to a focus plane in standard light microscopy. Thus, a stack of various amounts of two-dimensional reconstruction planes derived from a single hologram provides information of the complete observation volume.

The reconstruction algorithm is based on the calculation of the Fresnel-Kirchhoff Integral [13, 23]

$$K(\vec{r}) = \iint_S I\left(\vec{\xi}\right) \exp\left\{-\frac{i2\pi}{\lambda} \frac{\vec{\xi}}{\xi}\right\} d\vec{\xi} \tag{2.7}$$

where $\vec{r} = (x, y, z)$ is the position vector indicating a point in the observation plane, $\vec{\xi} = (\xi, \eta, L)$ denotes the coordinates on the screen at a distance L to the point source, λ is the wavelength, and $I\left(\vec{\xi}\right)$ is the intensity pattern of the hologram. The integral extends over the surface S of the recording screen. The Eq. (2.7) is only valid for a spherical source wave and under the assumption of the validity of the Fraunhofer condition [24]. The absolute value of $|K(\vec{r})|$ corresponds to the intensity distribution $I(x, y, z)$ in the reconstructed xy-layers along the z-axis.

In contrast to a photographic plate a CCD chip is discontinuous and consists of discrete pixels. Therefore the coordinates in the Eq. (2.7) have to be expressed as a discrete grid. Due to the non-linearity in the phase vector, Eq. (2.7) is extremely time-consuming to calculate. Kreuzer succeeded to develop an algo-rithm which is capable of removing the non-linearity in Eq. (2.7) and allows an exact and fast calculation [13]. More details on the algorithm can be found in the patent of Kreuzer [13] or in the Ph.D. thesis of Barth [1].

2.2.4 Resolution

In analogy to standard light microscopy, the resolution in digital in-line holog-raphy (DIH) is determined by the numerical aperture (NA) and the used wave-length. In general the NA is defined as

$$NA := n \cdot \sin \alpha \tag{2.8}$$

where n is the refractive index of the medium (1.0 for vacuum, 1.33 for pure water) and α is the half-angle of beam spread (see Fig. 2.2).

If the used wavelength and the pinhole diameter are of the same order in dimension the achieved resolution is not dependent on the pinhole diameter. This case is achievable if visible light is used as a source wave. In DIH the numerical aperture is determined by the pinhole-detector distance (L, see Fig. 2.2) and the size of the detector (D) if the latter is completely illuminated. Otherwise only the illuminated fraction of the screen has to be taken into account. By using visible light as a source wave it is possible to illuminate the complete detector. The numerical aperture in DIH is given by

$$NA = \frac{\frac{D}{2}}{\sqrt{L^2 + \left(\frac{D}{2}\right)^2}}. \tag{2.9}$$

Since $L \gg D$ the approximation $\sqrt{L^2 + \left(\frac{D}{2}\right)^2} \cong L$ is valid. Therefore the NA in DIHM is given as

$$NA = \frac{\frac{D}{2}}{L}. \tag{2.10}$$

The achievable resolution in DIH is subject to many studies and a detailed explanation can be found in Jericho et al. [16], Garcia-Sucerquia et al. [25] and Barth [1]. For the theoretical lateral and depth resolution follows

$$\delta_{\text{lateral}} = \frac{\lambda}{NA} \tag{2.11}$$

$$\delta_{\text{depth}} = \frac{\lambda}{2(NA)^2}. \tag{2.12}$$

According to those equations the achievable resolution for the setup used in the course of this thesis is

$$\delta_{\text{lateral}} = 2.3 \ \mu m \text{ and } \delta_{\text{depth}} = 5.6 \ \mu m. \tag{2.13}$$

References

1. R. Barth, Digital in-line x-ray holographic microscopy with synchrotron radiation, Ph.D. Dissertation, Ruprecht-Karls University of Heidelberg, Heidelberg, 2008
2. M. Schürmann, Digital in-line holographic microscopy with various wavelengths and point sources applied to static and fluidic specimens, Ph.D. Dissertation, Ruprecht-Karls-University of Heidelberg, Heidelberg, 2007

3. T. Gorniak, Digitale Rntgenholografie mit Lochblenden und Fresnelschen Zonenplatten bei BESSY und FLASH, Diploma Thesis, Ruprechts-Karls-University of Heidelberg, Heidelberg, 2009
4. P. Hariharan, *Basics of Holography* (Cambridge University Press, 2002)
5. U. Schnars, W. Jueptner, *Digital Holography* (Springer, Berlin, 2005)
6. E. Hecht, *Optik*, 4th edn. (Oldenbourg, 2005)
7. M.V. Klein, T.E. Furtak, *Optik*, 25th edn. (Springer-Verlag, 1988)
8. D. Meschede, *Optik, Licht und Laser*, 2nd edn. (Teubner, 2005)
9. H. Niedrig (ed.) *Bergmann-Schaefer: Lehrbuch der Experimentalphysik: Optik*, vol. 3, 10th edn. (Walter de Gruyter, 2004)
10. B.E.A. Saleh, M.C. Teich, *Fundamentals of Photonics*, 2nd edn. (Wiley, 2007)
11. W. Demtröder, *Experimentalphysik 2, Elektrizität und Optik*, 4th edn. (Springer, Berlin, 2006)
12. D. Gabor, Nature **161**(8), 777 (1948)
13. H.J. Kreuzer, US Patent 6.411.406, 2002
14. J. Garcia-Sucerquia, W. Xu, S.K. Jericho, M.H. Jericho, I. Tamblyn, H.J. Kreuzer, *Proc. SPIE Int. Soc. Optical Eng.*, 2006, 6026, 602613/1–602613/9
15. W. Xu, M.H. Jericho, H.J. Kreuzer, I.A. Meinertzhagen, Opt. Lett. **28**(3), 164–166 (2003)
16. S.K. Jericho, J. Garcia-Sucerquia, W.B. Xu, M.H. Jericho, H.J. Kreuzer, Rev. Sci. Instrum. **77**(4) 8 pp (2006)
17. M. Heydt, A. Rosenhahn, M. Grunze, M. Pettitt, M.E. Callow, J.A. Callow, J. Adhes. **83**(5), 417–430 (2007)
18. A. Rosenhahn, R. Barth, F. Staier, T. Simpson, S. Mittler, S. Eisebitt, M. Grunze, J. Opt. Soc. Am. A Opt. Image Sci. Vis. **25**(2), 416–422 (2008)
19. M. Malek, D. Allano, S. Coetmellec, C. Ozkul, D. Lebrun, Meas. Sci. Technol. **15**(4), 699–705 (2004)
20. J. Sheng, E. Malkiel, J. Katz, J. Adolf, R. Belas, A.R. Place, Proc. Natl. Acad. Sci. USA **104**(44), 17512–17517 (2007)
21. H. Sun, P.W. Benzie, N. Burns, D.C. Hendry, M.A. Player, J. Watson, Philos. Trans. R. Soc. A Math. Phys. Eng. Sci. **366**(1871), 1789–1806 (2008)
22. H.Y. Sun, B. Song, H.P. Dong, B. Reid, M.A. Player, J. Watson, M. Zhao, J. Biomed. Opt. **13**(1) 9 pp (2008)
23. H.J. Kreuzer, M.J. Jericho, I.A. Meinertzhagen, W.B. Xu, J. Phys. Condens. Matter **13**(47), 10729–10741 (2001)
24. Wikipedia, Web Page, Fraunhofer Diffraction, http://en.wikipedia.org/wiki/Fraunhofer_diffraction. Accessed 19 Nov 2009
25. J. Garcia-Sucerquia, W.B. Xu, S.K. Jericho, P. Klages, M.H. Jericho, H.J. Kreuzer, Appl. Opt. **45**(5), 836–850 (2006)

Chapter 3
State of the Art

The aim of this thesis is to study the exploration behavior of microorganisms and to correlate the observed behavior with the known antifouling performance of surfaces. In the following, the existing knowledge in literature important for the scope of this thesis is reviewed. First, the used organism is described followed by a detailed description of surface cues altering the observed settlement. Afterwards the general swimming properties of microorganisms are explained in detail. Finally the results of this thesis are compared to other 3D motion studies published in literature.

3.1 Alga *Ulva linza*

Within AMBIO [1], model organisms which represent the major fouling groups (microfouler, soft- and hard-macrofouler) are selected to study the performance of potential antifouling coatings. During the course of this thesis, experiments were done exclusively with spores of the green alga *Ulva linza* which represents the soft-macrofouling species.

Ulva commonly known as seaweed or sea lettuce is a bright green plant and can also be found in brackish water, particularly estuaries. The plant lives attached to an object (e.g. rocks, ship hulls, piers,…) in the middle to low intertidal zone. It is also found in a great amount on man-made structures, especially ship hulls, causes severe economical problems [2].

The life cycle of *Ulva* is shown in Fig. 3.1. All *Ulva* species are isomorphic, meaning that they alternate between gametophytic and sporophytic life stages with similar morphologies. In the center of Fig. 3.1, these life stages are illustrated by macroscopically visible plant leaves. The gametophytes are haploid and the sporophytes are diploid. The gametophytes produce biflagellate haploid gametes through mitosis. Male and (slightly larger) female gametes are positive phototactic and swim until they find each other and fuse. The formation of the syngamy is

M. Heydt, *How Do Spores Select Where to Settle?*, Springer Theses,
DOI: 10.1007/978-3-642-17217-5_3, © Springer-Verlag Berlin Heidelberg 2011

studied by video microscopy [3]. After the fusion, a negative phototatic spore is formed which swims towards a surface where it selects a place to settle and grow. Sporophytes produce quadriflagellate haploid zoospores through meiosis. These spores are negative phototactic and swim directly towards a surface to find a place to settle. This stage is regarded as the important step in reproduction of *Ulva* and therefore needs to be studied in detail to understand the mechanisms of surface colonization. Consequently, all experiments within the course of this thesis are done with spores. In Fig. 3.2a, a false colored scanning electron microscopy (SEM) image of a spore is shown.

The chemotaxis and the light stimulus response of *Ulva* spores is studied by Wheeler et al. [4]. The fruiting pattern of *Ulva* is controlled by the lunar cycle and the release of spores follows the gametes a few days later. The periodicity of gamete formation and release helps to ensure genetic exchange within the

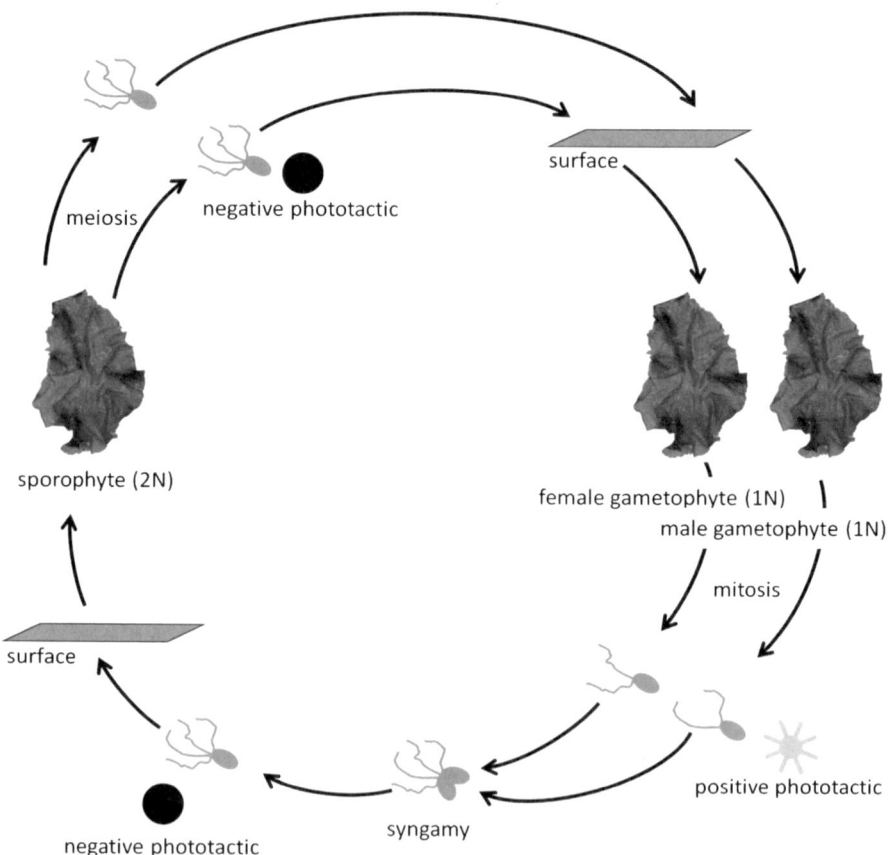

Fig. 3.1 *Ulva* life cycle

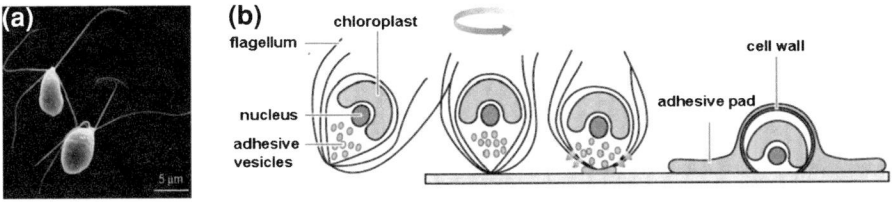

Fig. 3.2 **a** False colored SEM picture of Zoospores [6]; **b** Cartoon showing the steps involved in the settlement of *Ulva* spores [6]

population and is additionally assisted by the fact that each fond is of a different sex and therefore cannot self-fertilize.

The process of settlement has been studied in detail and is shown as a cartoon in Fig. 3.2 [5]. After a spore has selected a position on the surface a cue triggers the irreversible commitment to adhesion. Prior to and during the discharge of the contents of the adhesive vesicles (EPS) the spore rotates over a specific surface position [5]. During the adhesive release the spore withdraws the flagellar axonemes into the cell. This process (release, withdrawal) takes about 30–60 s and is followed by the synthesis of a new cell wall [5].

The adhesive is composed of polysaccharides and proteoglycans. In detail, it is a polydisperse, self-aggregating hydrophilic glycoprotein [5]. It is similar to the group of hydroxyproline rich extra cellular matrices of both plants and animals [5]. After the release the adhesives swells around 300 times and forms a pad around the spore. The latter starts to cure immediately after discharge. After the spore is firmly attached to the surface it starts to grow into a plant of macroscopically visible size (see life cycle in Fig. 3.1).

The motility and the flagellar beating pattern of various algae is studied by Inouye et al. [7]. To our knowledge the flagellar beat of *Ulva* zoospores has not yet been studied in detail by video microscopy. However, the swimming pattern of *Ulva* gametes has been studied, but gametes only have two flagella. Therefore Fig. 3.3 displays the flagellar beat pattern of other related quadriflagellate algae which, according to the literature, swim in a similar fashion [7]. During forward swimming these algae swing their flagella back along the cell body with the tips pointing backwards (see Fig. 3.3a). The beat pattern can be described by an undulatory wave produced at the base propagated towards the tips of the flagella. Figure 3.3b shows the flagella during forward swimming of *Prasinopapilla vascuolata*. The direction of the beat is radial and the four flagella are arranged in a cruciate pattern (Fig. 3.3c).

For *Ulva* gametes (Fig. 3.3h) only forward swimming is observed whereas for *Prasinopapilla vascuolata* (Fig. 3.3e) an avoiding response is observed in which the four flagella are in front of the cell. For the quadriflagellate *Cymbomonas tetramitiformis* (Fig. 3.3f, g) a backwards swimming pattern is observed. According to Inouye et al. [7]. *Cymbomonas tetramitiformis* is so far the only green algae examined which uses forward and backward swimming as normal

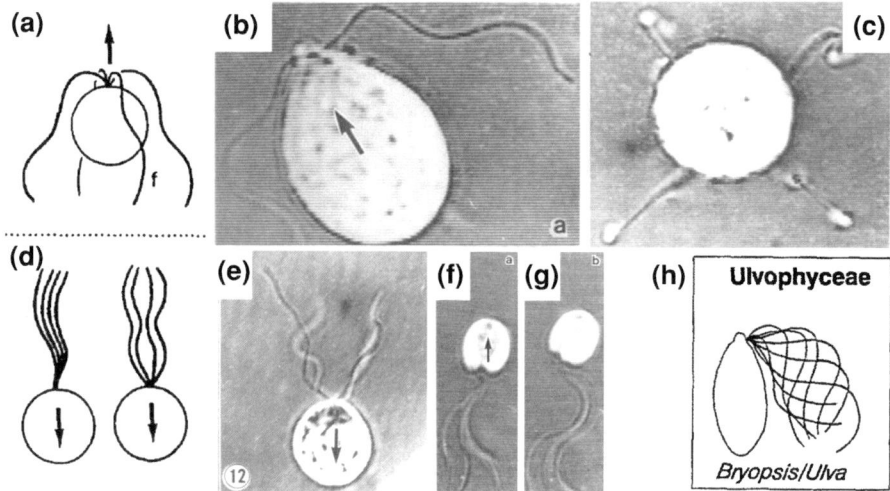

Fig. 3.3 a Schematic flagellar beta pattern for swimming "normal" (forward) of green algae of the genus: *Bryopsis, Claulerpa, Ulva, Cymobomonas* and *Prasinopapilla*; b One frame out of a video microscopy study showing the flagellar beat of *Prasinopapilla* vascuolata; c *Prasinopapilla* vascuolata cell view from the posterior side, showing cruciate profile of the flagellar beat. d Schematic flagellar beat pattern for the avoiding response of *Prasinopapilla* vascuolata and the backwards swimming flagellar beat pattern of *Cymobomonas tetramitiformis*; e Avoiding response of *Prasinopapilla* vascuolata; f backwards swimming flagellar beat pattern of *Cymbomonas tetramitiformis*; g backwards swimming flagellar beat pattern of *Cymbomona tetramitiformis*; h schematic flagellar beat pattern of Ulva (*Bryopsis maxina, Pertusa* and *Caulerpa raceemosa*) gamets while forward swimming; All images are taken form Inouye et al. [7]

methods of swimming. The forward and the backward swimming are reversible and spontaneously, but forward swimming appears to be more frequent. The forward swimming is similar to the motion observed for *Prasinopapilla vascuolat*. During the backwards swimming the four flagella are held in front of the cell and beat synchronously and uni-directionally (Fig. 3.3f, g).

The flagella movement plays an important role in the fusion of *Ulva* gametes which is similar to that of *Chlamydomonas* [3]. When a male gamete finds a female gamete, initial cytoplasmic contact between the two occurs at the tip of the flagella and the anterior end of the flagella base. Subsequently the male gamete establishes contact with the anterior end of the female gamete cell body, while for a short time period the contact between the flagella is lost. Following the distal part of the male flagellum comes in contact with that of the female gamete again. Subsequently, the male gamete maintains contact with the female gamete and the anterior side of the cell bodies and the flagella tips. The adhesion of the flagella tip is strong enough to hold during the ongoing flagella motion. Finally, the gametes lay side-by-side with their longitudinal axes nearly parallel to each other and fuse to form a syngamy.

3.2 Influence of Surface Properties on Settlement and the Adhesion Strength of *Ulva* Spores

Ulva zoospores are able to select their desired settling position on a surface. In the following different surface cues are presented which alter this selection. The standard way to study the latter is by simply counting how many spores of a known spore number have settled on the surface after a certain period of time.

The strength of spore attachment to the surface is dependent on the surface properties. Typically the adhesion strength is determined by shear water stress and the performance of the coating is commonly stated in percentage removal of the settled spores before and after the applied flow [8]. Surface parameters causing low settlement do not a priori imply weak spore surface attachment. In fact the opposite trend is observed for silicone elastomeric coatings based on poly(dimethylsiloxane) (PDMS). On these surfaces a high spore settlement is generally observed but the attached strength is small enough to allow washing the spores off the surface easily (e.g. caused by a moving vessel) [9]. Since, these coatings do not fulfill all necessary requirements to be named "antifouling coatings", yet inhibit permanent fouling, they are called "fouling release coatings". To this day, they are the only environmentally friendly coatings on the market.

3.2.1 Wettability

The influence of wettability is studied with various systems. On mixed OH/CH$_3$ alkanethiol self assembled monolayers (SAMs) on gold the spores avoid to settle on the hydrophilic surface and instead select the hydrophobic surface as shown in Fig. 3.4a [10–12]. However, although in the experiment more cells settled on the hydrophobic CH$_3$-enriched SAMs, they were weaker adhered than to the hydrophilic surfaces.

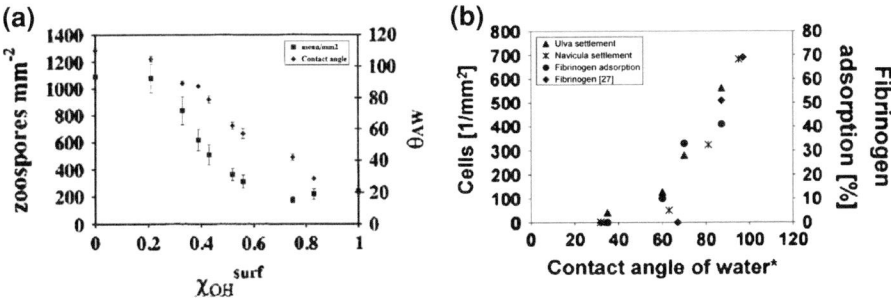

Fig. 3.4 Influence of wettability on spore settlement. **a** mixed OH/CH3-SAMs [12] and **b** EG$_6$-SAMs with different end groups [13]

On hexa(ethylene glycol)-containing SAMs (EG_6) with systematically changed end group [R-OH (contact angle (CA): 35°), R-OMe (CA: 60°), R-OEt (CA: 70°) and R-OProp (CA: 87°)] a similar trend is observed for the firmly adhered cells (see Fig. 3.4b) [13]. This is in good agreement to the protein resistance of the surface. If these surfaces can be studied in detail and the spores are counted on the surfaces, without passing the surfaces through the air–water interface, a completely different settlement trend is observed. The highest amount of settlement is observed on the protein resistant, hydrophilic EG_6-OH surface. Still, the spores are so weakly attached to the surface that by passing the air–water interface all spores are washed off the surface. Therefore, the expressed low settlement amount in Fig. 3.4b is not due to settlement inhibition per se, but rather to the extremely weak spore attachment to the surface and the removal of the spores from the surface during the handling of the sample during the assay.

In another study the influence on spore settlement was investigated on a patterned hydrophobic and hydrophilic surface [14]. Stripe pattern consisting of poly(ethylene glycol) (PEG) and fluorinated silane forming SAMs on silicon wafers were manufactured in different stripe width. In Fig. 3.5 the spore settlement is shown. Up to a stripe width of 5 µm the spores nearly exclusively settle on the

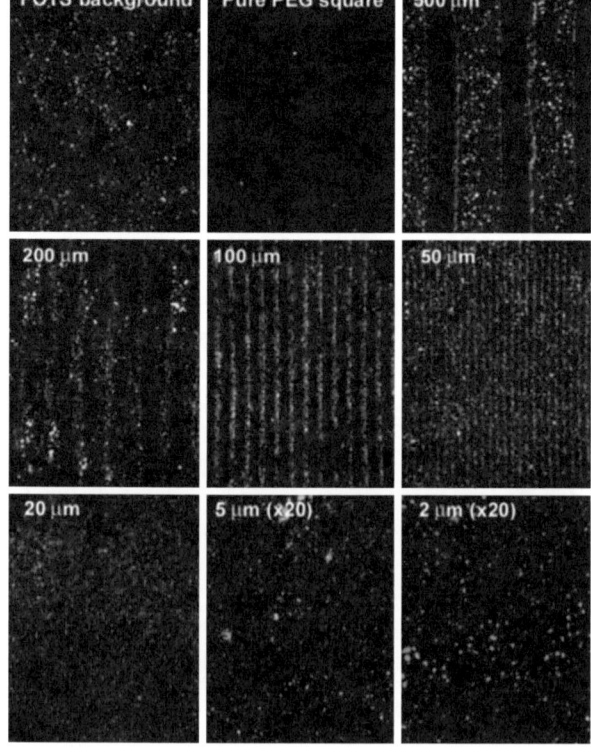

Fig. 3.5 Spore settlement on the PEGylated and fluorinated areas. The spores settle on the fluorinated strips exclusively up to a dimension of 5 µm. For 2 µm strips the spore settlement is equally distributed. Images are shown at a 10× magnification except the 5 and 2 µm images, which are shown at 20× magnification [14]

fluorinated areas whereas on the PEGylated areas hardly any spore settlement was observed. Thinner stripes (2 µm) were not recognized by the spores. In consequence, the spores settled homogeneously. However, up to a certain distance the spore can actively discriminate between the different surface chemistries.

The typical trend most observed (high settlement on hydrophilic and low on hydrophobic surface area) [11, 12] was confirmed for the homogeneous surfaces chemistries study by Chaudhury et al. [15]. However, if wettability gradients were generated out of the same components the amount of settlement is inverted [15]. In this publication the authors suggest that the gradient has a direct and active influence on the spore sensing abilities. This can be explained by the biased migration of the spores during the initial stages of surface sensing.

3.2.2 Ethylene Glycol Containing Surfaces Coatings

Poly(ethylene glycol)surfaces (PEG) are used in the course of this thesis. Ethylene glycol surfaces with different repeat units have been used for years in biomedical research as model surfaces in order to study their interaction with proteins, bacteria and cells [9, 16–19]. In the marine environment PEG surfaces are also studied [9, 19]. As long as the surface is stable under the experimental conditions the spores only settle in very small amounts on the surface and the very few attached spores are extremely weakly attached. Interestingly, the spore behavior on the EG_6-OH [13] is significantly different than on PEG [20] although the difference is just based on the amount of ethylene glycol repeat units (43 for PEG and 6 for EG_6). For both surfaces the contact angle, the surface energy and the observed settlement after the standard assay is similar. However, as already described in Sect. 3.2.1, on EG_6 the initial settlement during the assay is extremely high, but the spores are only loosely attached to the surface and are removed from the surface by handling the sample. In contrast this high initial settlement is not observed for PEG.

3.2.3 Lubricity

For the release rate the surface lubricity on a nano scale plays an important role whereas the settlement amount is not affected by lubricity [21]. Bowen et al. studied this effect by preparing alkanethiol SAMs of different chain lengths (C_8–C_{18}) [21]. The SAM structure changes from amorphous to crystalline but the surface energy and contact angle is similar for all prepared SAM. The change in structure is correlated to a higher release (see Fig. 3.6) and therefore depends on the friction coefficient of the surface.

Fig. 3.6 Percentage of
removal of cells dependent on
the friction coefficient of the
surface

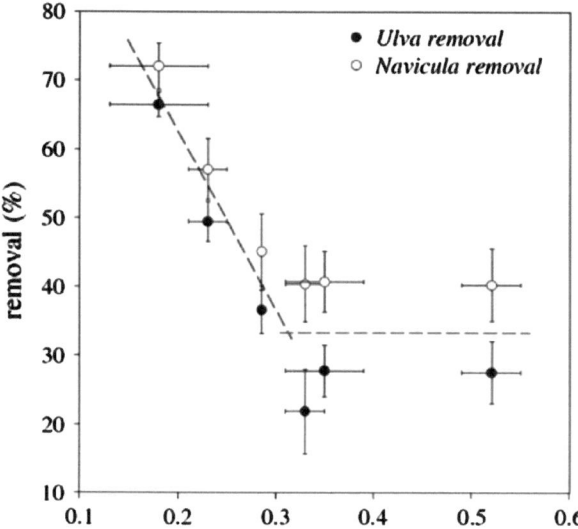

Fig. 3.6 Percentage of removal of cells dependent on the friction coefficient of the surface

3.2.4 Charge

The influence of a net surface charge on the settlement and adhesion of *Ulva* spores was studied by Rosenhahn et al. [22]. The zeta potential of a spore has been determined to be -19.3 ± 1 mV and a reduced tendency for spores to settle on negatively charged surfaces has been observed. If spores did settle on the latter, their adhesion strength was lower than on neutral or positively charged surfaces.

3.2.5 Topography

Spore settlement is strongly influenced by the topography of the surface [23–26]. If the topography provides depressions (e.g. holes, channels) which are large enough for the spores to fit inside, the surface is more attractive for settlement than a smooth surface of the same material [23–25]. If the feature size is smaller than the spore body (<5 μm) the surface is less attractive for settlement than a smooth surface [26–29]. If ridges and channels are arranged as shown in Fig. 3.7 the resulting pattern is called sharklet AF^{TM}. This surface structure is bio-inspired by the texture of the shark skin. The observed settlement on the sharklet AF^{TM} pattern is very low [28, 29]. The performance of the sharklet AF^{TM} is even better than other structured surfaces (pillar arrays, channels, triangles, …) with a feature size smaller than the spore body [28, 29]. Therefore the settlement inhibition has to be due to the specific arrangement of the features in the sharklet AF^{TM} pattern. The reason for the latter is under discussion [28].

Fig. 3.7 SEM image of the engineered Sharklet AF^{TM} pattern [29]

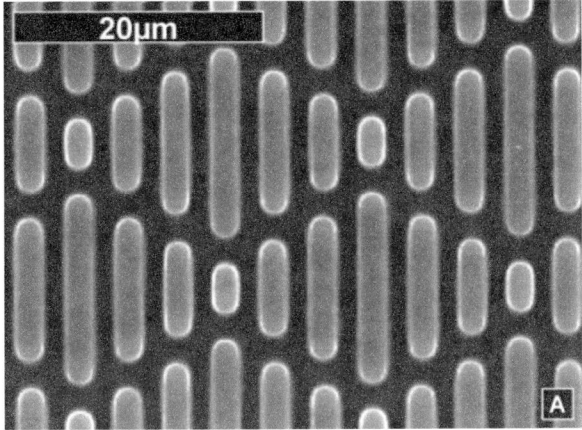

The sharklet AF^{TM} with a feature size (ridge width, channel width) smaller than 5 µm works very well against fouling of *Ulva* spores the surface, but fails as an antifouling surface in the real world. The reason for this is that the surface is encountered by a number of different species such as bacteria, algae, and barnacles ranging several orders of magnitude (from hundreds of nanometers to centimeters). To prevent fouling by topography hierarchically structures are necessary [30, 31]. Schumacher et al. showed that is possible to prevent both barnacles settlement and *Ulva* spore settlement if the sharklet AF^{TM} pattern is superimposed on a larger channel structure [29]. Recently the performance of hierarchical structures to prevent biofouling has been studied in a field test [32]. The surface remains relatively free from biofouling even after prolonged exposure to seawater for 18 months.

3.3 Motility of Microorganisms

The fluid dynamics of fast, large self-propelled objects ranking from krill to whales is extensively studied [33, 34], but is fundamental different to the motility of microorganisms [35]. For these large objects the Reynolds number (Re) ranges from somewhat >1 to enormous (2×10^8 for a whale [36]). For microscopic swimmers (bacteria, uni- and multi-cellular algae and protists) which, although capable of swimming many body lengths per seconds, Re ≪ 1 [37, 38]. Although their presence is usually not immediately obvious, microorganisms constitute the major part of the world's biomass. The fluid habitants of these organisms range from soil moisture and melting snow to lakes, oceans and even saturated saline ponds. Their population dynamics—replication and decay, accumulation and dispersal—modulates and regulates their own life, the life of larger creatures that feed on them, and even the climate [39]. The motion of peritrichously flagellated bacteria, e.g. *Escherichia coli (E. coli)*, close to surfaces are relevant to understanding

the early stage of biofilm formation and of pathogenic infections [40–43]. Such
motion next to solid boundaries is different from the motility in boundary-free
solution for many microorganisms, as it has most extensively been studied for *E.
coli* [44–46].

3.3.1 Hydrodynamics Basics: Life at Low Reynolds Number [37]

The Reynolds number (Re) indicates a measure for the ratio of inertial forces to
viscous forces. For a typical microorganism the Reynolds number ranges from of
10^{-4} to 10^{-5} and can be calculated according to Eq. 3.1 where η is the viscosity of
the medium, a the diameter of the object, ρ the fluid density, v the velocity of the
object and υ the kinematic viscosity (10^{-2} cm^2/s for water). At small Re the
inertial forces can be neglected for a swimming microorganism and the swimming
performance is dominated by the viscous forces.

$$\text{Re} := \frac{\text{inertialforces}}{\text{viscous forces}} \approx \frac{av\rho}{\eta} = \frac{av}{\upsilon} \qquad (3.1)$$

The significance of low Reynolds number for the motility is clarified in the
following example. A bacterium swims with a speed of 30 μm/s in average. If a
bacterium is pushed by an external force to move and suddenly the force vanishes,
it will coast within 0.1Å and about 0.6 μs. Therefore a microorganism's action at a
specific moment is determined by the forces on the organism at this moment only
and not by earlier exerted forces [47]. The energy *E. coli* has to spend in order to
move can be calculated to 2×10^{-9} μW assuming the efficiency of the swimming
apparatus to be 1%.

3.3.2 Properties of Swimming Microorganisms

The group of swimming microorganisms includes bacteria, spermatozoa and other
gametes, unicellular and colonial algae and protozoa. Their mean diameter lies
between 1 and 200 μm. These organisms are denser than the water in which they
swim, by a few percent for algae, approximately 10% for bacteria such as B.
subtitlis, and 30% for spermatozoa [48].

Microorganisms propel themselves through the water by using waving, undu-
lating or rotating appendages called flagella or cilia which are arranged in various
geometries [49]. A typical flagellum has a diameter of 45–130 Å where a cilia has
a diameter of about 1,000 Å [37]. The swimming speeds (v_s) can reach up to
several hundred μm/s. One intensively studied plant cell system are algae from the
genus *Chlamydomonas* ($d = 10$–20 μm, $v_s = 50$–100 μm/s). These organisms are
in approximation a spheroid that swims by a breaststroke like motion of two

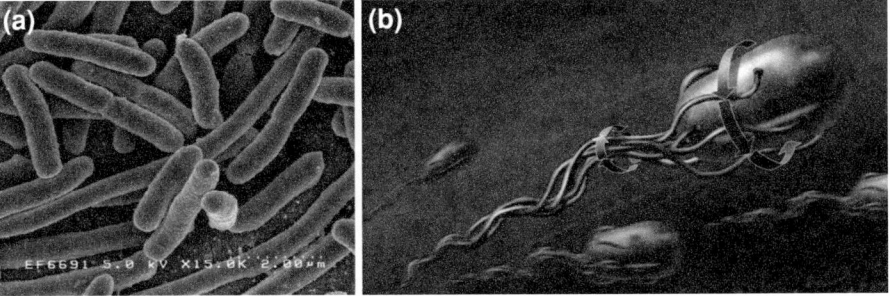

Fig. 3.8 a SEM picture of *E. coli*; **b** model for the "run" swimming phase of *E. coli*. The flagellum bundle is rotating CCW while the cell body is rotating CW [55]

flagella, which are attached near the anterior end [50–54]. For this organism the mass distribution is typically anisotropic, so that the center of mass is posterior to the center of buoyancy.

Another well studied microorganism is *E. coli*. *E. coli* is propelled from the rear by several rotation flagella [44]. Figure 3.8b shows a semantic view of *E. coli*. The motor (flagellum) is able to rotate clockwise (CW), as seen by an observer standing outside of the cell looking down at the hook, or counterclockwise (CCW).

When the motor turns CCW, the filaments rotate parallel in a bundle that pushes the cell body steadily (in a straight-line) forward. This movement is called "run". While the motor rotates CCW and the cell moves forward the cell body has to turn CW to balance external torque [56]. When the motor turns CW, the flagella turn independently and the cell body moves erratically with little net displacement. This movement is described as "tumble". The two modes alternate. The cell runs and tumbles, executing a three-dimensional random walk, as shown in Fig. 3.9. The mean run interval is about 1 s, whereas the tumble interval is only 0.1 s.

The change in angle between the runs is generated by the tumble phase and is approximately random (with a slight forward bias). If a cell follows a special gradient of a chemical attractant the runs are extended. If the bacterium chooses

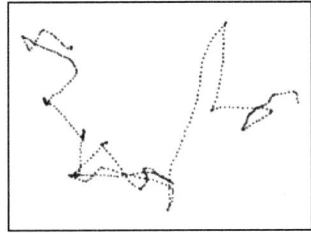

Fig. 3.9 This plot shows about 30 s in the life of an *E. coli* K-12 bacterium swimming in an isotropic homogenous medium. The track spans about 0.1 mm from left to right. The plot shows 26 runs and tumbles, the longest run (nearly vertical) lasting 6.6 s. The mean speed is 21 μm/s [57]

the opposite direction, runs revert to the length observed in the absence of a gradient. Thus, the cell is able to move up or down the gradients. The mechanism of gradient sensing of *E. coli* is temporal and not spatial. *E. coli* does not determine whether there is more attractant, say, in front than behind; rather, it determines whether the concentration increases when it moves in a particular direction. Studies of impulsive stimuli indicated that a cell compares the concentration observed over the past 1 s with the concentration observed over the previous 3 s and responds to the difference [58, 59]. The runs in Fig. 3.9 are not quite straight because the cell is subject to rotational Brownian movement that causes it to wander off course by about 30° within 1 s. After about 10 s it drifts off course by more than 90° and "forgets" the direction in which it was going.

3.3.3 Hydrodynamics Interaction at Solid Boundaries

The role of hydrodynamic interaction in the locomotion of microorganisms is theoretically described in a number of publications [37, 38, 49, 56, 60–65]. These publications explain the movement in solution away from any boundary. Rothschild [66] in the mid 1960s and Berke et al. [67] more recently studied the distribution of spermatozoa and *E. coli* between two glass slides (shown in Fig. 3.10). Although they used different organisms, both measured a very similar cell distribution.

The favored explanation for that result is the presence of hydrodynamic forces as described in a number of theoretical calculations [67–69]. In comparison to the

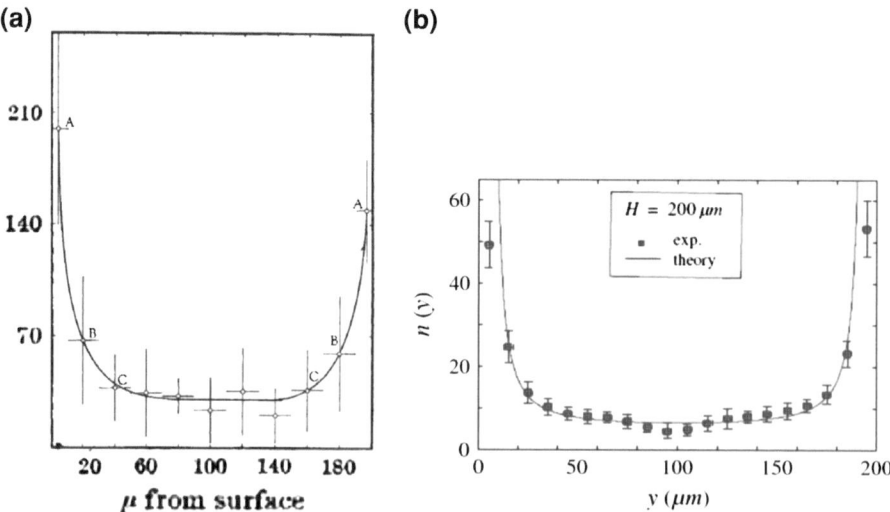

Fig. 3.10 Cell distribution between tow glass surfaces. **a** bull spermatozoa [66]; **b** *E. coli* [67]

motile organisms a non motile object, e.g. a polystyrene bead, is repelled by a surface due to hydrodynamic forces if no attractive forces (like e.g. electrostatics) are present [70]. Figure 3.11 shows this theoretical calculation for a 2.5 µm large particle approaching a surface. The drag force acting on a moving object in the low Reynolds number regime is described by Stokes' law and is dependent on the radius r of the object and its velocity. When an object approaches a solid interface the drag forces (Stokes' law, Eq. 3.2) has to be modified by a correction factor γ.

$$F = 6\pi r v \gamma \qquad (3.2)$$

The drag force increases at a distance of 10 µm from the surface. For swimming objects the hydrodynamic forces caused by the beating of the flagella have to be taken into account as well. In the case of swimming cells the repulsion by the drag force is overcompensated by different attractive forces. The flow around most flagellated swimming organisms, including spermatozoa cells or bacteria such as *E. coli*, is well approximated by a force dipole (stresslet) [48]. The flagellum provides the force which is opposed by the drag on the cell body. As the cell swims it sets up a dipolar flow field which in general does not satisfy the no-slip boundary conditions near a solid interface. The flow in the presence of the interface is the result of the linear superposition of the infinite-fluid, plus any image flow field on the other side of the surface which is necessary to enforce the boundary conditions near the solid interface [65]. Physically, the approach to image flow fields is similar to the method of images in electrostatics, with the only difference that in fluid dynamics the image is a vector field. This force dipole induces a flow of fluid towards the wall. To gain physical intuition, it is easier to picture a dipole near a free surface; in that case, the image system is an equal dipole on the other side of the surface, and two parallel dipoles attract each other [67]. For the motility of cell this flow results in a change of swimming direction into a movement parallel to the surface as schematically illustrated in Fig. 3.12.

Biondi et al. measured the velocity distribution of *E. coli* in micro channels [71]. They found that *E. coli* swims faster in a 3 µm height channel than in a free solution or 10 µm channel. Ramia et al. predict in their theoretical calculation that

Fig. 3.11 Theoretical calculation for the distance dependent increase of the drag force (γ) of an object of the same size as an *Ulva* spore [70]

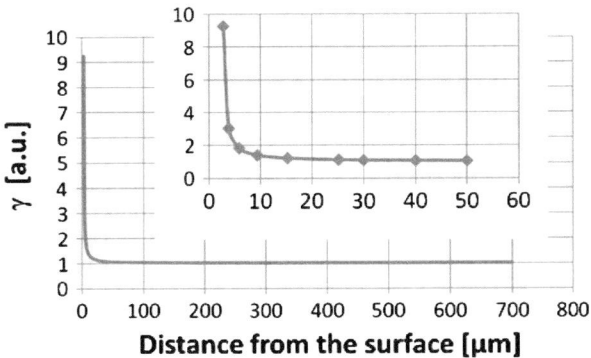

Fig. 3.12 Induced rotation of swimming cells by a solid interface. A cell which is pushed from behind is hydrodynamically orientated into a swimming direction parallel to the surface [65]

a back-propelled microorganism moves faster when it is swims in close vicinity to the surface [38]. This speed increase near a wall is a result of the increasing hydrodynamic forces near the solid interface. The reason for that is that the propulsive force of the flagellum is drag based and therefore more effective near the wall [65]. However, if the swimmer is assumed to swim with a constant power, then the boundary condition near the solid interface in general leads to a decrease of the swimming speed [65]. Magariyama et al. studied the motion pattern of the mono flagellated *Vibrio alginolyticus* in solution and near different chemical modified surfaces [72]. They observed a significant difference in the motion pattern if the cell is close to the surface in comparison to the motion in solution. Drescher et al. showed that the spherical algae *Volvox* are attracted to each other and can form stable bound states in which they "waltz" around each other if they swim near a solid boundary [73]. These bound states are induced by fluid dynamics and hydrodynamics near the surface [73].

If *E. coli* is closer than 2 μm to the surface, it does not swim in straight runs but is observed to trace out clockwise circles [46, 74]. The swimming direction is clockwise when viewed from above. The circular swimming motion of the cell is due to the rotation of the cell body and hydrodynamic drag. This effect has been used to predict the direction of cellular motion in the micro channel [45].

As described above, microorganisms can spend periods of times (>1 min) swimming in close proximity to a surface [74–76]. In irreversible adhesion, by contrast, bacteria adhering to the surface do not move, either by swimming or Brownian motion [40, 77, 78]. Before cells irreversible adhere, they can (but do not have to) become "tethered" to the surface. "Tethered" describes a state when a flagellum adheres to the surface but the cell body still rotates freely [79, 80]. "Tethered" cells are in an intermediate state between swimming and irreversible adhesion because they are able to either start swimming again or adhere to the surface. Adhesion of individual cells to a surface is the first step in the formation of biofilms [81]. The standard model for bacterial adhesion implies that bacteria proceed from a loosely attached, reversible state to a more strongly attached, irreversible state [82]. Bacteria within a biofilm are much harder to kill than swimming bacteria [83]; therefore it would be of advantage to control cellular adhesion and prevent the very formation of harmful biofilms [84].

3.4 Tracking in 3D

Tracking light microscopy has been used to investigate bacteria like *E. coli* [85]. By using this technique it is possible to track one bacterium at a time in three dimensions by using translation stages. It was found that a single bacterium switches its motility patterns over time between swimming and tumbling phases [86]. Another way of measuring three-dimensional trajectories is stereoscopy, which requires two synchronized cameras [87, 88]. Confocal microscopy has also been used to study the motion of particles in colloidal systems over time, however the nature of these scanning technique limits the achievable frame rate [89]. Another common technique to image a flow-profile is μPIV, which uses small tracer particles [90]. However, traditional PIV methods were found to have difficulties in mapping the complex three dimensional trajectories.

Digital in-line holography (DIH) provides an alternative: lensless microscopy technique which intrinsically contains complete three-dimensional information about the investigated volume. The methode does not require a feedback control which responds to the motion and it uses only one CCD camera. This very straightforward method and can be realized by a very simple setup as shown in Fig. 4.1a or explained in Sect. 2.2. Holography has been applied in the past to study swimming marine organisms such as algal spores [91–95], predator–prey interactions of dinoflagellates and the resulting changes in their behavior [96] and in microfluidics to visualize liquid flow fieled [97, 98].

References

1. EC Framework 6 Integrated Project "AMBIO"—Advanced Nanostructured Surfaces for the Control of Biofouling, http://www.ambio.bham.ac.uk. Accessed 23 Nov 2009
2. M.E. Callow, J.A. Callow, J.D. Pickett-Heaps, R. Wetherbee, J. Phycol. **33**(6), 938–947 (1997)
3. S. Miyamura, Cytologia **69**(2), 197–201 (2004)
4. G.L. Wheeler, K. Tait, A. Taylor, C. Brownlee, I. Joint, Plant Cell Environ. **29**(4), 608–618 (2006)
5. J.A. Callow, M.E. Callow, *Biofilms*, 1st edn. (Springer-Verlag, Berlin, Heidelberg, 2006)
6. M.E. Callow, J.A. Callow, Biologist **49**(1), 1–5 (2002)
7. I. Inouye, T. Hori, Protoplasma **164**(1–3), 54–69 (1991)
8. M.P. Schultz, J.A. Finlay, M.E. Callow, J.A. Callow, Biofouling **15**(4), 243–251 (2000)
9. A. Statz, J. Finlay, J. Dalsin, M. Callow, J.A. Callow, P.B. Messersmith, Biofouling **22**(6), 391–399 (2006)
10. L.K. Ista, M.E. Callow, J.A. Finlay, S.E. Coleman, A.C. Nolasco, R.H. Simons, J.A. Callow, G.P. Lopez, Appl. Environ. Microbiol. **70**(7), 4151–4157 (2004)
11. J.A. Finlay, M.E. Callow, L.K. Ista, G.P. Lopez, J.A. Callow, Integr. Comp. Biol. **42**(6), 1116–1122 (2002)
12. M.E. Callow, J.A. Callow, L.K. Ista, S.E. Coleman, A.C. Nolasco, G.P. Lopez, Appl. Environ. Microbiol. **66**(8), 3249–3254 (2000)
13. S. Schilp, A. Kueller, A. Rosenhahn, M. Grunze, M.E. Pettitt, M.E. Callow, J.A. Callow, Biointerphases **2**(4), 143–150 (2007)

14. J.A. Finlay, S. Krishnan, M.E. Callow, J.A. Callow, R. Dong, N. Asgill, K. Wong, E.J. Kramer, C.K. Ober, Langmuir **24**(2), 503–510 (2008)
15. M.K. Chaudhury, S. Daniel, M.E. Callow, J.A. Callow, J.A. Finlay, Biointerphases **1**(1), 18–21 (2006)
16. M. Mrksich, G.M. Whitesides, Annu. Rev. Biophys. Biomol. Struct. **25**, 55–78 (1996)
17. J.H. Harris, *Poly (Ethylene Glycol) Chemistry: Biotechnical and Biomedical Applications* (Plenum Press, New York, 1992)
18. S. Herrwerth, W. Eck, S. Reinhardt, M. Grunze, J. Am. Chem. Soc. **125**(31), 9359–9366 (2003)
19. S. Schilp, A. Rosenhahn, M.E. Pettitt, J. Bowen, M.E. Callow, J.A. Callow, M. Grunze, Langmuir **25**(17), 10077–10082 (2009)
20. S. Schilp, Self-assembled Monolayers and Nanostructured Surfaces as Tools to Design Antifouling Surfaces, Ph. D. Dissertation, Ruprecht-Karls-University of Heidelberg, Heidelberg, 2009
21. J. Bowen, M.E. Pettitt, K. Kendall, G.J. Leggett, J.A. Preece, M.E. Callow, J.A. Callow, J. R. Soc. Interface **4**(14), 473–477 (2007)
22. A. Rosenhahn, J.A. Finlay, M.E. Pettit, A. Ward, W. Wirges, R. Gerhard, M.E. Callow, M. Grunze, J.A. Callow, Biointerphases **4**(1), 7–11 (2009)
23. L. Hoipkemeier-Wilson, J.F. Schumacher, M.L. Carman, A. Gibson, A. Feinberg, M.E. Callow, J.A. Finlay, J.A. Callow, A.B. Brennan, Biofouling **20**(1), 53–63 (2004)
24. M.E. Callow, A.R. Jennings, A.B. Brennan, C.E. Seegert, A. Gibson, L. Wilson, A. Feinberg, R. Baney, J.A. Callow, Biofouling **18**(3), 237–245 (2002)
25. W.R. Wilkerson, C.A. Seegert, A.W. Feinberg, L.C. Zhao, J.A. Callow, M.E. Callow, A.B. Brennan, Polymer Pre. **42**(1), 147–148 (2001)
26. A.J. Scardino, J. Guenther, R. de Nys, Biofouling **24**(1), 45–53 (2008)
27. M.L. Carman, T.G. Estes, A.W. Feinberg, J.F. Schumacher, W. Wilkerson, L.H. Wilson, M.E. Callow, J.A. Callow, A.B. Brennan, Biofouling **22**(1), 11–21 (2006)
28. J.F. Schumacher, C.J. Long, M.E. Callow, J.A. Finlay, J.A. Callow, A.B. Brennan, Langmuir **24**(9), 4931–4937 (2008)
29. J.F. Schumacher, M.L. Carman, T.G. Estes, A.W. Feinberg, L.H. Wilson, M.E. Callow, J.A. Callow, J.A. Finlay, A.B. Brennan, Biofouling **23**(1), 55–62 (2007)
30. J. Genzer, K. Efimenko, Biofouling **22**(5), 339–360 (2006)
31. A.J. Scardino, D. Hudleston, Z. Peng, N.A. Paul, R. de Nys, Biofouling **25**(1), 83–93 (2009)
32. K. Efimenko, J.A. Finlay, M.E. Callow, J.A. Callow, J. Genzer, ACS Appl. Mater. Interfaces **1**(5), 1031–1040 (2009)
33. M.H. Dickinson, C.T. Farley, R.J. Full, M.A.R. Koehl, R. Kram, S. Lehman, Science **288**(5463), 100–106 (2000)
34. J.M.V. Rayner, J. Exp. Biol. **80**(1), 17–54 (1979)
35. J.T. Bonner, Nat Hist **115**(9), 54–59 (2006)
36. K.A. Kermack, Exp. Biol. **25**(3), 237–240 (1948)
37. E.M. Purcell, Am. J. Phys. **45**(1), 3–11 (1977)
38. M. Ramia, D.L. Tullock, N. Phanthien, Biophys. J. **65**(2), 755–778 (1993)
39. R.J. Charlson, J.E. Lovelock, M.O. Andreae, S.G. Warren, Nature **326**(6114), 655–661 (1987)
40. M.A.S. Vigeant, R.M. Ford, M. Wagner, L.K. Tamm, Appl. Environ. Microbiol. **68**(6), 2794–2801 (2002)
41. R.M. Harshey, Annu. Rev. Microbiol. **57**, 249–273 (2003)
42. K.M. Ottemann, J.F. Miller, Mol. Microbiol. **24**(6), 1109–1117 (1997)
43. L.A. Pratt, R. Kolter, Mol. Microbiol. **30**(2), 285–293 (1998)
44. H.C. Berg, D.A. Brown, Nature **239**, 500 (1972)
45. W.R. DiLuzio, L. Turner, M. Mayer, P. Garstecki, D.B. Weibel, H.C. Berg, G.M. Whitesides, Nature **435**(7046), 1271–1274 (2005)
46. E. Lauga, W.R. DiLuzio, G.M. Whitesides, H.A. Stone, Biophys. J. **90**(2), 400–412 (2006)
47. A. Franklin, Am. J. Phys. **44**(6), 529–545 (1976)

48. T.J. Pedley, J.O. Kessler, Annu. Rev. Fluid Mech. **24**, 313–358 (1992)
49. C. Brennen, H. Winet, Annu. Rev. Fluid Mech. **9**, 339–398 (1977)
50. U. Rüffler, W. Nultsch, Cell Motil. Cytoskeleton **5**(3), 251–263 (1985)
51. U. Rueffer, W. Nultsch, Cell Motil. Cytoskeleton **7**(1), 87–93 (1987)
52. U. Rüffer, W. Nultsch, Bot. Acta. **108**(3), 255–265 (1995)
53. K.A. Johnson, Bioessays **17**(10), 847–854 (1995)
54. C.D. Silflow, P.A. Lefebvre, Plant Physiol. **127**(4), 1500–1507 (2001)
55. Wikipedia, Web Page, *Escherichia coli*, http://en.wikipedia.org/wiki/Ecoli. Accessed 12/2/2008
56. E.M. Purcell, Proc. Natl. Acad. Sci. USA **94**(21), 11307–11311 (1997)
57. H.C. Berg, Phys. Today **53**(1), 24–29 (2000)
58. J.E. Segal, S.M. Block, H.C. Berg, Proc. Natl. Acad. Sci. USA **83**, 8987–8991 (1986)
59. H.C. Berg, Annu. Rev. Biochem. **72**, 19–54 (2003)
60. w. Ludwig, J. Comp. Physiol. A Neuroethol. Sens. Neural. Behav. Physiol. **13**(3), 397–504 (1930)
61. J. Lighthill, SIAM Rev. **18**(2), 161–230 (1976)
62. J. Lighthill, J. Eng. Math. **30**(1–2), 25–34 (1996)
63. J. Lighthill, J. Eng. Math. **30**(1–2), 35–78 (1996)
64. J.E. Avron, O. Gat, O. Kenneth, Phys. Rev. Lett. **93**(18), 4 (2004)
65. E. Lauga, T.R. Powers, Rep. Prog. Phys. **72**(9), 36 (2009)
66. L. Rothschild, Nature **198**(4886), 1221–1222 (1963)
67. A.P. Berke, L. Turner, H.C. Berg, E. Lauga, Phys. Rev. Lett. **101**, 3–038102(1)–(4) (2008)
68. J.P. Hernandez-Ortiz, C.G. Stoltz, M.D. Graham, Phys. Rev. Lett. **95**(20), 204501 (2005)
69. L.J. Fauci, A. McDonald, Bull. Math. Biol. **57**(5), 679–699 (1995)
70. H. Brenner, Chem. Eng. Sci. **16**(3–4), 143–339 (1961)
71. S.A. Biondi, J.A. Quinn, H. Goldfine, AIChE J. **44**(8), 1923–1929 (1998)
72. Y. Magariyama, M. Ichiba, K. Nakata, K. Baba, T. Ohtani, S. Kudo, T. Goto, Biophys. J. **88**(5), 3648–3658 (2005)
73. K. Drescher, K.C. Leptos, I. Tuval, T. Ishikawa, T.J. Pedley, R.E. Goldstein, Phys. Rev. Lett. **102**(16), 168101 (2009)
74. P.D. Frymier, R.M. Ford, H.C. Berg, P.T. Cummings, Proc. Natl. Acad. Sci. USA **92**(13), 6195–6199 (1995)
75. H.C. Berg, L. Turner, Biophys. J. **58**(4), 919–930 (1990)
76. M.A.S. Vigeant, R.M. Ford, Appl. Environ. Microbiol. **63**(9), 3474–3479 (1997)
77. J. Lyklema, W. Norde, A.J.B. Zehnder, Microb. Ecol. **17**(1), 1–15 (1989)
78. J. Palmer, S. Flint, J. Brooks, J. Ind. Microbiol. Biotechnol. **34**(9), 577–588 (2007)
79. S.H. Larsen, R.W. Reader, E.N. Kort, W. Tso, J. Adler, Nature **249**(5452), 74–77 (1974)
80. M. Silverman, M. Simon, Nature **294**(5452), 73–74 (1974)
81. M.C.M. van Loosdrecht, J. Lyklema, W. Norde, A.J.B. Zehnder, Microbiol. Rev. **54**(1), 75–87 (1990)
82. L. Marcotte, A. Tabrizian, IRBM **29**(2–3), 77–88 (2008)
83. R. Srinivasan, P.S. Stewart, T. Griebe, C.I. Chen, X.M. Xu, Biotechnol. Bioeng. **46**(6), 553–560 (1995)
84. S.Y. Hou, E.A. Burton, K.A. Simon, D. Blodgett, Y.Y. Luk, D.C. Ren, Appl. Environ. Microbiol. **73**(13), 4300–4307 (2007)
85. H.C. Berg, Rev. Sci. Instrum. **42**(6), 869–871 (1971)
86. H.C. Berg, *Random Walks in Biology* (Princeton University Press, Princeton, 1993)
87. S.A. Baba, S. Inomata, M. Ooya, Y. Mogami, A. Izumikurotani, Rev. Sci. Instrum. **62**(2), 540–541 (1991)
88. K. Drescher, K.C. Leptos, R.E. Goldstein, Rev. Sci. Instrum. **80**(1), 7 (2009)
89. E.R. Weeks, J.C. Crocker, A.C. Levitt, A. Schofield, D.A. Weitz, Science **287**(5453), 627–631 (2000)
90. J.G. Santiago, S.T. Wereley, C.D. Meinhart, D.J. Beebe, R.J. Adrian, Exp. Fluids **25**(4), 316–319 (1998)

91. W. Xu, M.H. Jericho, H.J. Kreuzer, I.A. Meinertzhagen, Opt. Lett. **28**(3), 164–166 (2003)
92. N.I. Lewis, W.B. Xu, S.K. Jericho, H.J. Kreuzer, M.H. Jericho, A.D. Cembella, Phycologia **45**(1), 61–70 (2006)
93. M. Heydt, A. Rosenhahn, M. Grunze, M. Pettitt, M.E. Callow, J.A. Callow, J. Adhes. **83**(5), 417–430 (2007)
94. H.Y. Sun, D.C. Hendry, M.A. Player, J. Watson, IEEE J. Ocean. Eng. **32**(2), 373–382 (2007)
95. S.K. Jericho, J. Garcia-Sucerquia, W.B. Xu, M.H. Jericho, H.J. Kreuzer, Rev. Sci. Instrum. **77**, 4–43706 (2006)
96. J. Sheng, E. Malkiel, J. Katz, J. Adolf, R. Belas, A.R. Place, Proc. Natl. Acad. Sci. USA **104**(44), 17512–17517 (2007)
97. J. Sheng, E. Malkiel, J. Katz, Exp. Fluids **45**(6), 1023–1035 (2008)
98. J. Garcia-Sucerquia, W. Xu, S.K. Jericho, M.H. Jericho, H.J. Kreuzer, Optik **119**(9), 419–423 (2008)

Chapter 4
Experimental Details

In this chapter the developed hardware (holographic microscope, wet cell, heat isolation, subsonic noise isolation) and software (data acquisition, reconstruction, trace determination and trace interpretation) is described in detail. Furthermore, all experimental details regarding surface position determination, preparation of the used organisms and the synthesis of the used surfaces are provided.

4.1 Setup

4.1.1 Holographic Microscope

A transportable holographic microscope was developed in order to be able to measure in different labs. The basic idea for the instrument follows the implementation described earlier by Xu et al. [1]. In Fig. 4.1a, a picture of the main components is shown. Laser, focusing optics, pinhole, wet cell (including the sample) and CCD are set up in the in-line geometry.

The optical components are aligned horizontally whereas the wet cell for the sample is mounted vertically in order to prevent any debris from sinking down to the surface driven by gravity. Objective (Euromex Microscopes, 20×, NA 0.4, The Netherlands) and pinhole (500 nm, National Apertures, USA) are mounted in a cage system (Thorlabs GmbH, Germany). The objective can be moved within the cage system independently in respect to the pinhole by an adjustable platform (change in z-position). This adjustment is necessary to perfectly focus the laser onto the pinhole and thereby maximize the photon flux on the CCD-Chip. The pinhole is fixed at a position in the cage system but mounted on a xy translator (travel distance 13 mm, Thorlabs GmbH, Germany) to move it precisely into the focus of the beam. A DPSS laser (green, 532 nm, 10 mW, IMM Meßtechnologie, Germany) is coupled into the optical axis via two mirrors. Two irides are used in

M. Heydt, *How Do Spores Select Where to Settle?*, Springer Theses,
DOI: 10.1007/978-3-642-17217-5_4, © Springer-Verlag Berlin Heidelberg 2011

(a)   **(b)**

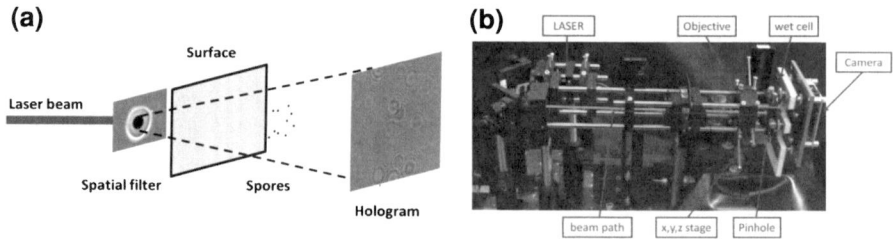

Fig. 4.1 a Schematic sketch of the in-line geometry; **b** picture of the built in-line holograph. The optical path is horizontal aligned in a cage system (Thorlabs GmbH, Germany)

the cage system to align the laser beam and to clean the beam profile. The irides also block unwanted reflections from the objective and adjusting mirrors.

The CCD-camera is a Lumenera Lu160M with 1,392 × 1,040 pixels (1,280 × 1,024 active pixels) and an area of 6.45 × 6.45 μm/pixel as applied previously for underwater digital in-line holography [2]. The sensor is also mounted on the cage system which allows adjusting the distance between pinhole and CCD-camera continuously. While performing the experiments, the distance between the chip and the pinhole was typically 16 mm, which is a compromise of optimal illumination of the complete CCD chip, large enough magnification and big enough observation volume. For the camera used, the best achieved frame rate was 11 Hz.

The wet cell is not attached to the cage system to keep as much flexibility as possible and to allow a simple surface exchange. The wet cell is mounted on an individual xyz adjustable platform in order to have three degrees of freedom to access a random position of choice and to optimize the magnification. To obtain the necessary resolution to resolve *Ulva* spores the wet cell pinhole distance has to be smaller than 1 mm.

To make the instrument deployable in different environments the microscope is shielded from daylight by a wooden box. The wooden protection device is equipped with windows to access the translation stages during the measurement. Those windows are covered with laser curtains to protect the setup from stray light. Including the wooden box the instrument has a dimension of 52 × 32 × 40 cm, weighs 6 kg and is therefore easily transportable.

4.1.2 Wet Cell

In Fig. 4.2 the observation chamber is shown. The cell is designed for fast exchange of the surface being investigated. The chamber consists of three parts, the observation chamber itself made from Teflon with a volume of 1 mm^3 and two steel lids. The lids are used to seal the cell by pressing a transparent surface (e.g. a glass cover slip) on the o-ring embedded in the observation chamber. The lids on each side can be changed independently. This setup allows an easy exchange of

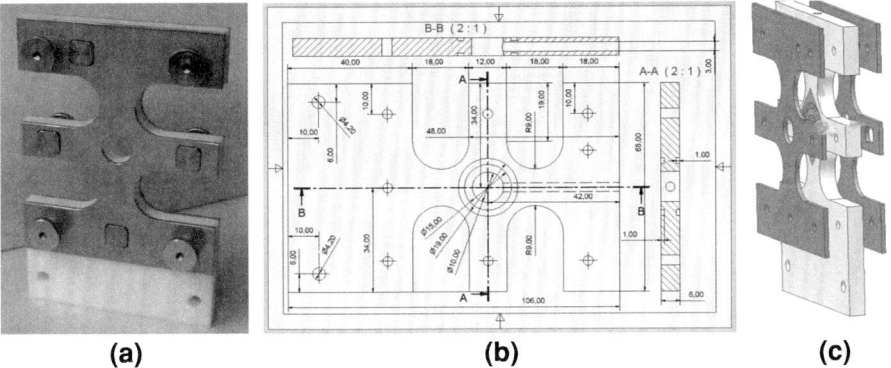

Fig. 4.2 Wet cell: **a** CAD assembly; **b** drawing with dimensions; **c** complete assembled wet cell

sealing surfaces. In Fig. 4.2b a drawing with dimensions is shown to clarify the real size of the cell. Panel c shows the finished and complete assembled wet cell. The surfaces used to seal the wet cell can be used simultaneously, e.g. as samples to study the exploration behaviour of swimmers in vicinity to a surface.

4.1.3 Subsonic Noise Isolation

The setup is isolated from subsonic noise by the use of four posts of soft tissue situated underneath the instrument.

4.1.4 Data Acquisition Program

For the exact calculation of the spore velocities the holograms were stored with a timestamp, as the frame rate for acquiring the images is not constant. Figure 4.3a shows an example for a trajectory recorded with inconstant frame rate. In the magnified area (see panel a) of the trajectory the spore suddenly moves approximately five times the distance it usually moves. In Fig. 4.3b the velocity is calculated for the trajectory shown in panel a. When the velocity is calculated assuming a constant frame rate (gray curve in panel b) the velocity jumps for a single point to a high value. This velocity value is not correct.

The data is acquired via an USB 2.0 port so that the hard disk speed limits the used frame rate. The storage of the images can cause a delay for further acquired images. To overcome this problem a timestamp with a precision in the range of milliseconds was included in the filename for each acquired image. The modification of the existing Labview® program was done with the help of Florian Staier, a former colleague from the work group. The time information is provided by the

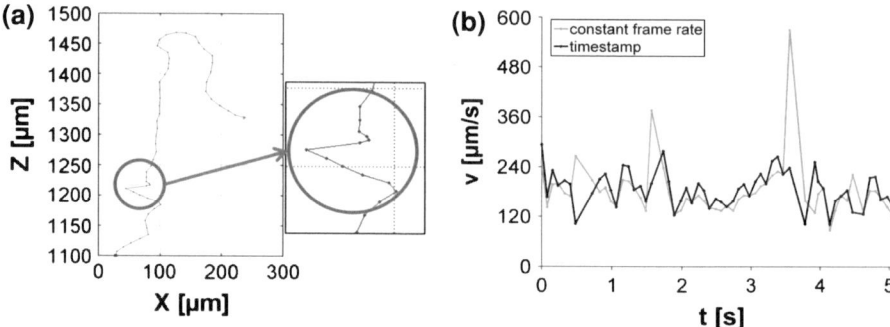

Fig. 4.3 Example for the recording without a constant frame rate. **a** xz view of a spore movement with magnification of the readout delay; **b** velocity calculation for the spore trajectory shown in **a**

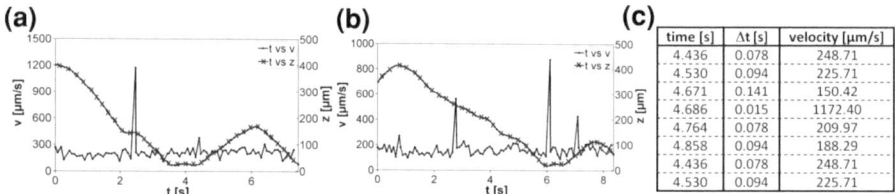

Fig. 4.4 Two examples for an incorrectly stored timestamp. In **a**, **b** the velocity is plotted on the left side (*no marker*) and the change in z-position is plotted on the right side (*marker*: x). Both are plotted against the elapsed observation time. In **c** the details for the spike in **a** are shown

Windows system time. In addition a faster hard disk was bought to keep the readout delays as small as possible.

In the course of the detailed motion analysis it turned out that the timestamp is not always stored accurately. Two examples in which the velocity jumps for a single data point to an extremely high value are shown in Fig. 4.4. In Fig. 4.4a at 2.3 s the velocity jumps from 150 to 1,173 µm/s and back to 210.0 µm/s. In the z-position (marker: x, scale on the right side) and in the x- and y-position (not shown) no big changes are found which could explain the velocity spikes. A readout delay as described above can also be excluded. Therefore the error has to be due to the timestamp. In Fig. 4.4c the time values for the plot in panel a are shown. The velocity spike occurs in the fifths row where Δt is only 0.015 s.

To exclude a systematic jitter the timestamp was assigned to the previous (timestamp −1) or the next (timestamp +1) image. Even if the velocity is calculated with these new assignments the spikes in the velocity data still occur (see Fig. 4.5). A possibility to avoid the high values is to calculate the velocity assuming a constant frame rate. For the data shown in Fig. 4.4 this approach works well. But, as described above, the data is not actually stored with a constant frame

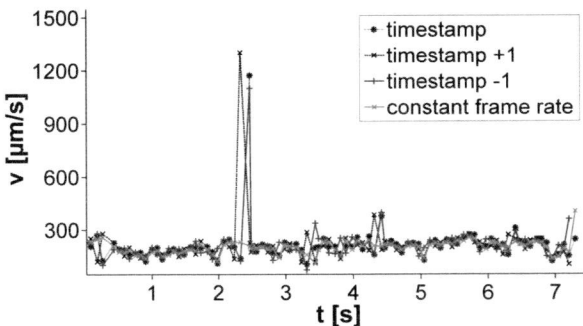

Fig. 4.5 Example for the timestamp jitter in the calculated velocity. Even if the saved timestamp is assigned to the previous ("timestamp −1") or the next ("timestamp +1") image the velocity spikes still occur. Only if the velocity is calculated assuming a constant frame rate the velocity jump can be prevented

rate and therefore this approach is not suitable for all data even if it works in this special case.

Even if the saved timestamp is not always stored correctly the data can still be analyzed more accurate than without the timestamp because the readout delay can be analyzed correctly. For the data recorded and analyzed within this thesis the timestamp is therefore used. To avoid errors in the interpretation of the data any velocity value greater as 500 µm/s is excluded.

For future experiments a new camera (pco. 1,200 s) will be used where the timestamp is provided by the camera software and is not added by Labview® when the image is stored on the hard disk. This camera also makes use of an individual storage and is therefore not prone to limited acquisition speed by the data transfer. The new camera was purchased not only because of the timestamp problem but as well for many other reasons such as higher frame rate (max. 636 fps), bigger CCD chip (12.3 × 12.3 mm), passive cooling of the CCD, and higher signal to noise ratio (SNR).

4.1.5 Heat Isolation

A major problem occurring in the first acquired data was that in addition to the cell's self-propulsion motion a convection induced flow was recorded. Figure 4.6 shows a typical example for the obtained data for *Ulva* spores with superimposed convectional flow.

The convection is caused by the waste heat of the CCD-chip. To solve this problem the CCD chip was cooled by a nitrogen stream. Therefore a rubber tube was installed in the instrument. To adjust the flow of nitrogen a temperature control was build in next to the CCD-chip to keep the temperature constant during the measurement. With this setup it was possible to record convection free data.

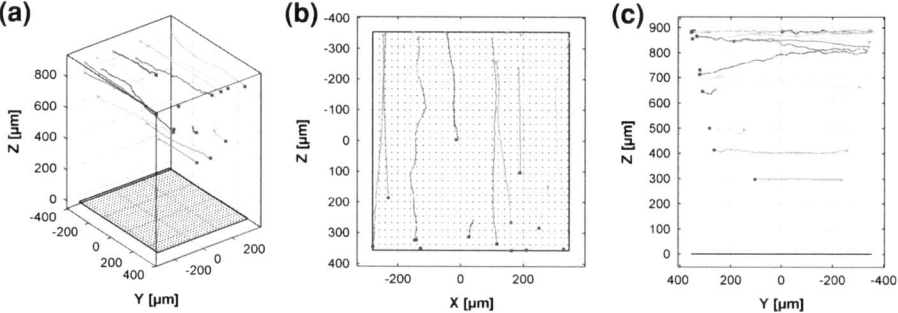

Fig. 4.6 Example for convectional flow **a** 3D view; **b** xy view; **c** xz view

4.2 Experimental Procedure

Figure 4.7 shows a flowchart of the experimental procedure. The measurement was done with the instrument described in Sect. 4.1 in the labs of Prof. Callow in Birmingham, UK. While running the experiments the first analysis "online filtering for events" is done. The analysis is basically done by watching the interference pattern recorded by the camera. The main focus of this analysis is to check the following:

- check for a stable illumination
- check whether convection occurs in the observation volume (see Sect. 4.1.5)
- check the swimming performance of the spores (dead or alive)
- note surface events (if possible to witness)

If the experiment was successful (based on the parameters above) the holograms are stored in a "Data bank A". The data volume recorded is in the range of 2 Gigabyte per experiment. Before the holograms are reconstructed the data is

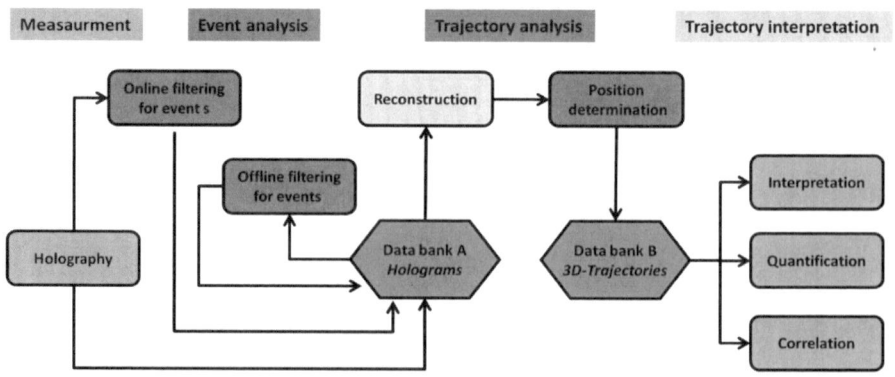

Fig. 4.7 Schematic overview of the procedure from the experiment to the motion analysis

analyzed again ("Offline filtering for events") to search for events of interest (for example settlement events). In Sect. 5.2.3 the analysis is shown in detail.

For the trajectory analysis and the trajectory interpretation individual programs are written in matlab® (Mathworks 2007a) which are briefly explained in the next Sects. 4.3 and 4.4).

4.3 Trajectory Analysis

In this section the trajectory analysis (see flowchart shown in Fig. 4.7) is explained. The section is divided into two parts: (1) reconstruction and (2) position determination.

4.3.1 Reconstruction

To analyze sequences of thousands of images a batch reconstruction program is needed because the existing reconstruction program only works for a single image. The reconstruction routine itself is described in detail in Sect. A.2.2. The structure of the program is kept extremely simple for a fast execution. It is based on a spread sheet, which the program executes line by line. In the spread sheet (saved as a text file) all needed parameters are defined.

Before reconstructing a complete frame sequence (500–1,000 frames), the reconstruction distances (L11 and L12) have to be determined. L11 and L12 define the position of the lowest (L11, nearest to the point source) and highest (L12, furthest to the point source) reconstruction plane. To analyze the exploration behaviour of zoospores on a surface the lower distance (L11) has to be below the surface. L11 is defined by reconstructing a few composite holograms (generated out of ≈ 20 images) at different points in time in the frame sequence. In this composite hologram the lowest object position is determined. From the obtained value for L11 100 µm is subtracted to make sure not to cut off any objects in the following reconstruction. The higher distance (L12) is typically 800–1,200 µm higher than L11 somewhere in the solution. After defining these parameters the resulting volume is reconstructed in steps of 5 µm. A step width of 5 µm was determined to be the perfect value to save reconstruction time but to be still able to determine the position very accurately. If the step width is smaller more time to reconstruct the images is necessary without achieving a better z-determination. In general the depth resolution for most experiments is anyway only 5.6 µm (see Eq. 2.13). For a 800 µm high volume this results in a stack of 160 reconstructed images for each hologram. If a large dataset is reconstructed the result is a huge amount of data which involves the problem of data handling similar to time lapse measurements in confocal microscopy. Therefore the need of storage capacity for the reconstructed data was reduced by generating three projections from the three

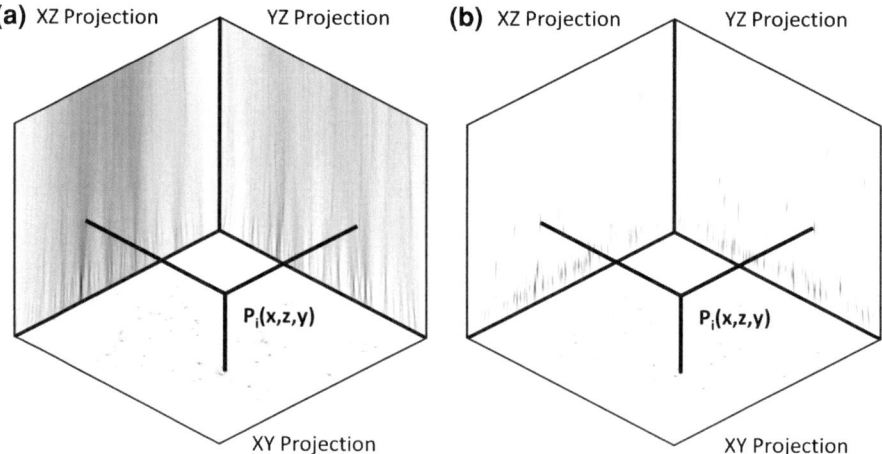

Fig. 4.8 Calculated projections from the reconstruction stack. **a** simple addition in XZ- and YZ-projection; **b** addition of only the max values in XZ- and YZ-projection [3]

dimensional image stacks. These are the XY-, XZ- and YZ-projections. In Fig. 4.8 typical projections are shown.

For the XY-projection a simple addition of each stack layer works very well, but for the XZ and YZ projection this leads to very noisy data (see Fig. 4.8a) and makes it impossible to reliably extract coordinates. Therefore each pixel in a XZ- and YZ-projection only contains the maximum value present in the projected rod of pixels. A typical result for the three projections obtained is shown in Fig. 4.8b. The low noise in the background allows for distinguishing the objects in the volume from the background. For the XY projection also a small improvement in the signal to noise (SNR) ratio is achieved by only using the maximum value present in the projection rod of pixels (Fig. 4.8b). These projections are stored on a hard disk before the position determination is started.

4.3.2 Position Determination

It turned out to be necessary to have a customized position determination program to be able to analyze many spores' positions fast and accurately. This data is needed to study the exploration behavior of *Ulva* spores. The position determination software is one main effort of the thesis. The history of the software development is summarized shortly.

The first approach was to use imageJ® with the plugin "object tracker". This determination works well in the xy-plane but to determine the z-coordinate a different program had to be used because the "object tracker" only determined in 2D. This approach worked well enough to analyze the first motion data and to obtain a few trajectories of moving spores. Yet in total this approach was too time consuming and too inaccurate in all spatial directions as it dependent on the position which the user chose by hand.

The second approach followed the idea of a completely automated position determination, which was implemented in Matlab® together with Peter Divós. But the routine turned out to be very error prone so that checking the data needed the same amount of time as determining the position using the first approach.

The third and last approach is a combination of the earlier ones. It is implemented in Matlab® and has a graphical interface to allow easy use, especially for other users. The program is implemented the way that the trajectories are analyzed subsequently. After a start position is manually assigned to a spore, the program automatically determines the spore positions in the consecutive frames until it leaves the field of view (FoV) or cannot be definitely allocated (e.g. crossing of trajectories).

To locate the 3D position of an *Ulva* spore in the projections, the position in the XY plane is first determined. After an initial coarse determination of the position by the user, an area with a size of only three times the spore diameter is used for locating the centre of mass of the spore. This is necessary to avoid problems caused by other spores in its vicinity. In order to disregard brighter tails or neighboring *Ulva* spores, a threshold is applied which sets all points with a brightness below a certain gray value to zero. If more than one object is present in the image cutout the algorithm keeps the object of interest and discards the other object. This threshold parameter usually lies in the order of 60% of the brightest feature and has to be optimized to the imaging conditions and to the contrast the objects causes. Subsequently the corresponding z-positions are determined from the XZ and YZ projection. Self-consistency needs to be maintained for the two projections. Using the starting point in the three spatial directions of the first frame in the time sequence, the subsequent frame is analyzed automatically in the same way under the assumption that the spore does not swim further than maximal four times the mean travel speed. The automatic routine runs until the spore cannot definitely be allocated or it is aborted by the user. If the spore density is small enough within only approximately 20 objects being present in one projection at a time, the determination works nearly without user intervention. In the end, uncertain position determinations (mainly lacking self-consistency in the z values) are marked and thus can manually be refined by the user. To realize this position determination program the important steps are shown in a flowchart (see Fig. 4.9).

The program is controlled by the main user interface (see flowchart). In this interface (see Fig. 4.10 for a screenshot) all necessary parameters for the determination can be adjusted. It turned out to be necessary to exclude points (e.g. static objects) which can interfere with the automated determination. Figure 4.11 shows a screenshot of this interface ("Exclusion points"). "The trajectory start point determination" has an individual interface (not shown as an individual figure) in which the start values for a trajectory can be defined. These values are added to a list which is automatically evaluated by the "automated trajectory determination"

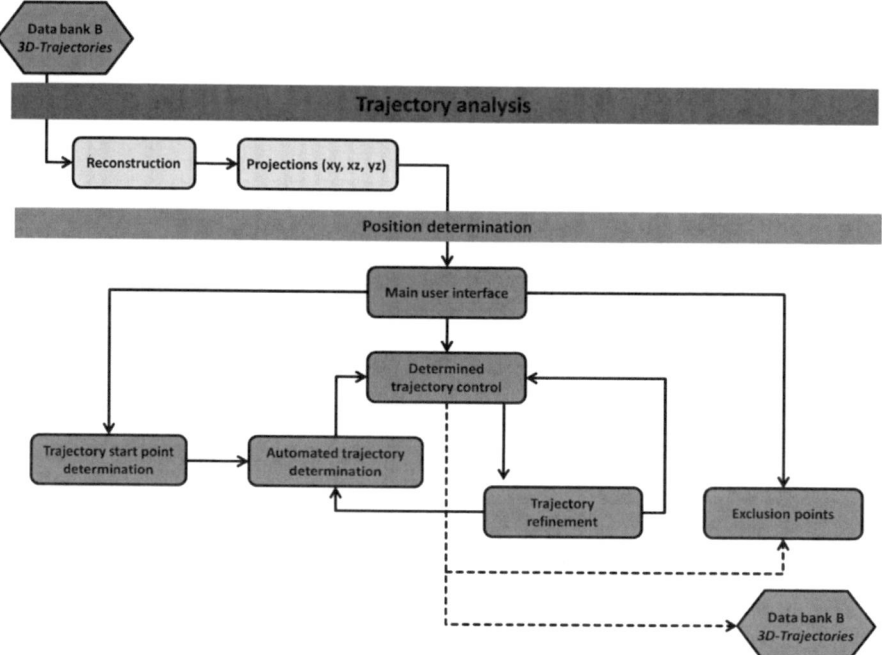

Fig. 4.9 Flowchart of the position determination program (see Fig. 4.8 for an complete overview of the experimental procedure)

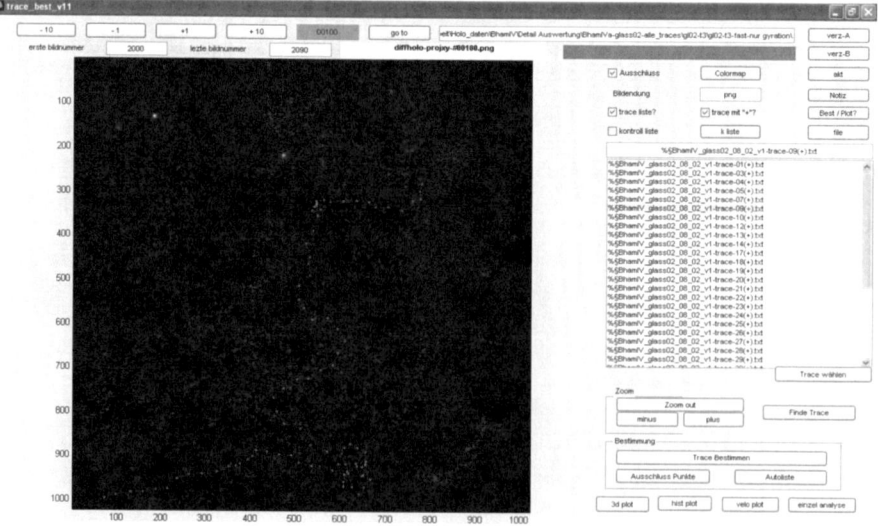

Fig. 4.10 Screenshot of the main user interface

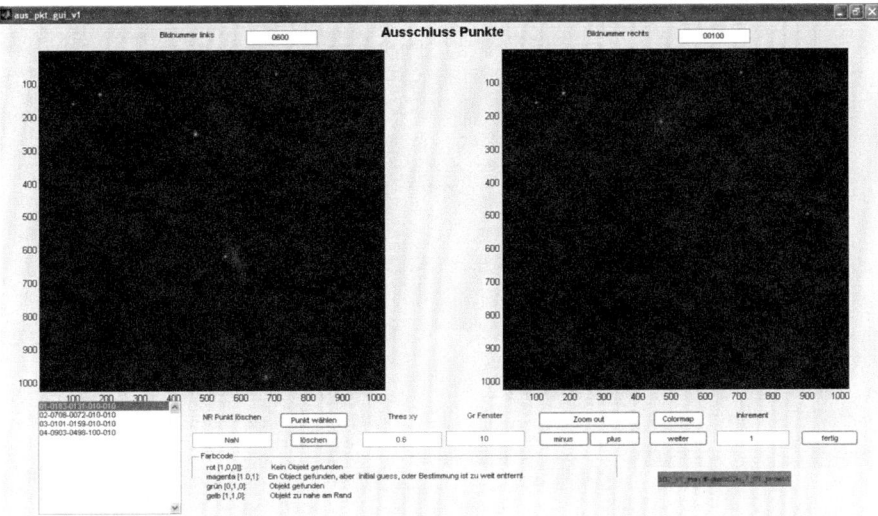

Fig. 4.11 Screenshot of the "exclusion points" interface

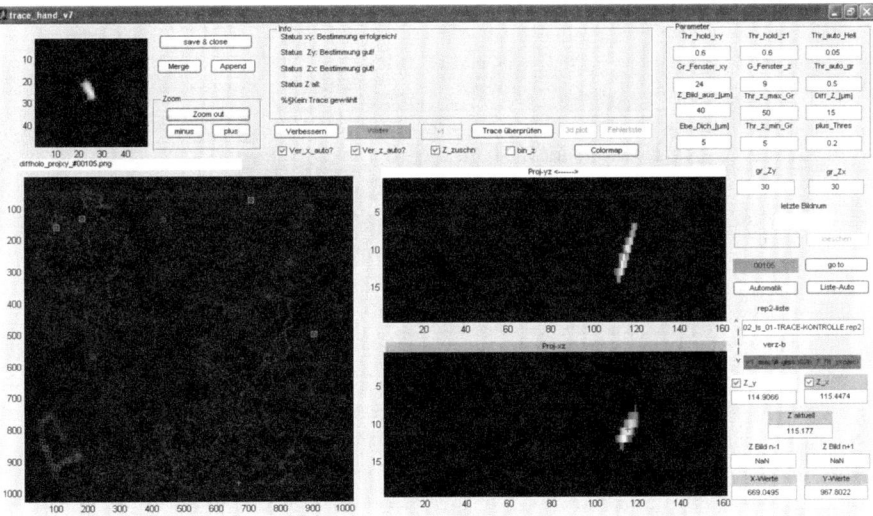

Fig. 4.12 Screenshot of the "determined trajectory control" interface

routine. If the automatic routine is not able to determine the next spore position, the trajectory is marked as "finished; wait for user responds" and is passed to the "Determined trajectory control" routine. The interface for this routine is shown in Fig. 4.12.

Before a trajectory can be added to the "Data bank B" it has to be checked by the user. The "Determined trajectory control" interface makes it easy to refine the trajectory, paste it back to the automated routine after solving a problem (e.g. crossing of two spores), or add the trajectory to the "Data bank B" when the trajectory is completely determined (the spore leaves the FoV or the last image is reached). If the trajectory is added to the data bank the spore positions of this trajectory are excluded from the automatic determination. This exclusion allows the routine to evaluate crossing points because some objects are known and cannot be determined anymore. Therefore the automatic determination works better with increasing amount of analyzed positions.

The strength of this algorithm is that it is programmed in a way that the automatic routine runs independently on a separate computer while the user can mark new start values or refine/check already automatic determined positions. The data between the two computers is always synchronized to allow a fast determination. All presented trajectories are analyzed with this software.

4.4 Trajectory Interpretation

The trajectories stored in "Data bank B" are used to analyze the spore exploration behavior (see flowchart in Fig. 4.7). To study the latter it was necessary to implement an analysis tool ("trace interpretation interface") with a graphical interface (see Fig. 4.13 for a screenshot of the interface). In the following the options of the analysis tool are described.

From the low numerical aperture in the used setup it becomes obvious that the z-resolution is worse compared to the x/y-resolution (see Eq. 2.13). This is illustrated in Fig. 4.14, where a static particle on the surface was imaged 500 times. In the determined positions the x-position remains quite stable and varies only by roughly 500 nm (approximately 15% of the spore diameter) but the z-determination varies by 2.5 µm, which is in the order of the diameter of the spore body. Although this uncertainty does not affect the descriptive component within the thesis, the analysis of velocity histograms will suffer as the z-component of the velocity vector contains more noise. Therefore, spans of 13 data points were approximated by a 2nd-degree polynomial model in a local regression and the different regressions were weighted by linear least squares while disregarding severe outliers. This results in a drastically reduced noise which is much closer to the one in the XY projections. This ensures that the width of the histogram is similarly influenced by the uncertainty present in all three vector components of the single positions. In the "trace interpretation interface" the span of the polynomial model can be altered to smooth the data according to the experimental needs because the resolution in x/y and z depends on several variables (see Eq. 2.13) and can therefore change between the experiments.

To understand the exploration behavior throughout the course of the thesis many different analysis approaches are implemented and combined in the "trace

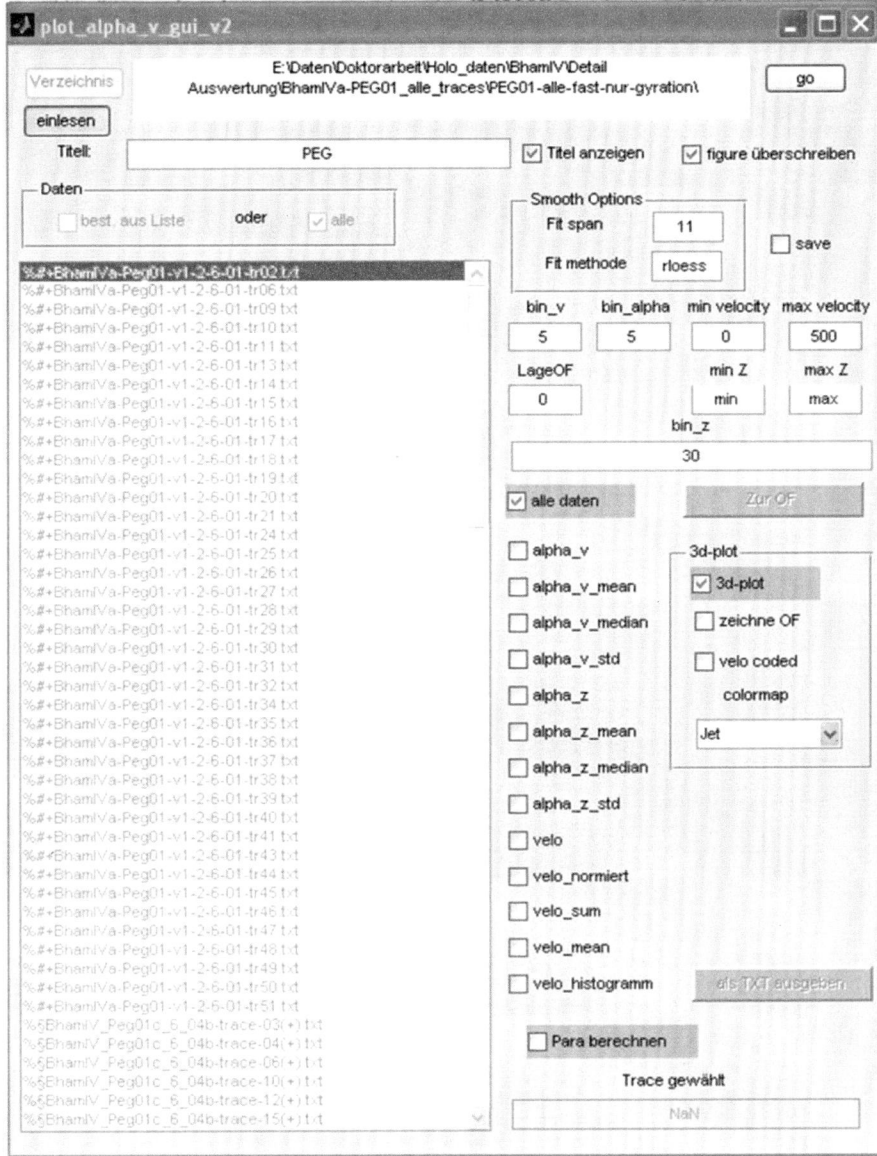

Fig. 4.13 Screenshot of trace interpretation interface

interpretation interface". In the following only the used analysis approaches are
listed and in the brackets an example within the thesis is shown:

- 3D plot for the trajectories
  - complete volume (Fig. 5.13)

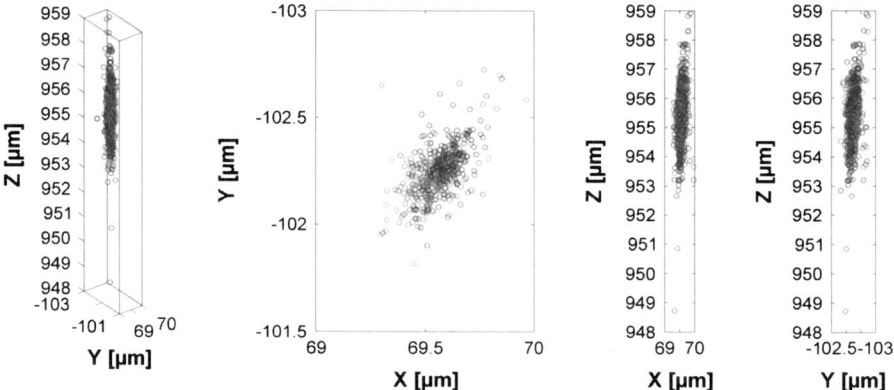

Fig. 4.14 Fluctuation in position determination for a static point on the surface recorded over 40 s [3]

- individual trajectories (Fig. A.69)
- color coded for different velocities (Fig. 5.7)

• velocity histogram

- individual trajectories (Fig. 5.5)
- section of the observation volume (Fig. A.79)

• angular (α_v and α_z) and velocity distribution

- individual trajectories (Fig. A.69)
- section of the observation volume (Figs. A.82 and A.83)

• distribution of mean values (mean velocity (v_m), $\bar{\alpha}_v$ and $\bar{\alpha}_z$)

- section of the observation volume (Fig. A.82 and Fig. A.83)
- detailed analysis to characterize the movement in the section towards and away from the surface (Fig. 6.3)

• detailed trajectory analysis

- velocity versus time (Fig. A.69)
- α_v versus time (Fig. A.69)
- α_z versus time (Fig. A.69)
- Distance from the surface versus time (Fig. A.69)

• spore density distribution (Fig. A.81)

The angle α_v (see Fig. 4.15a) is defined as the angle between two consecutive displacement vectors. If for example α_v is 0° the spore swims in a straight line and if α_v is 180° it swims backwards.

The angle α_z is defined as the angle of the spore velocity vector with respect to the surface normal and is illustrated in Fig. 4.15b. If α_z is smaller than 90° the

Fig. 4.15 Sketch to illustrate the definition of α_v (**a**) and α_z (**b**)

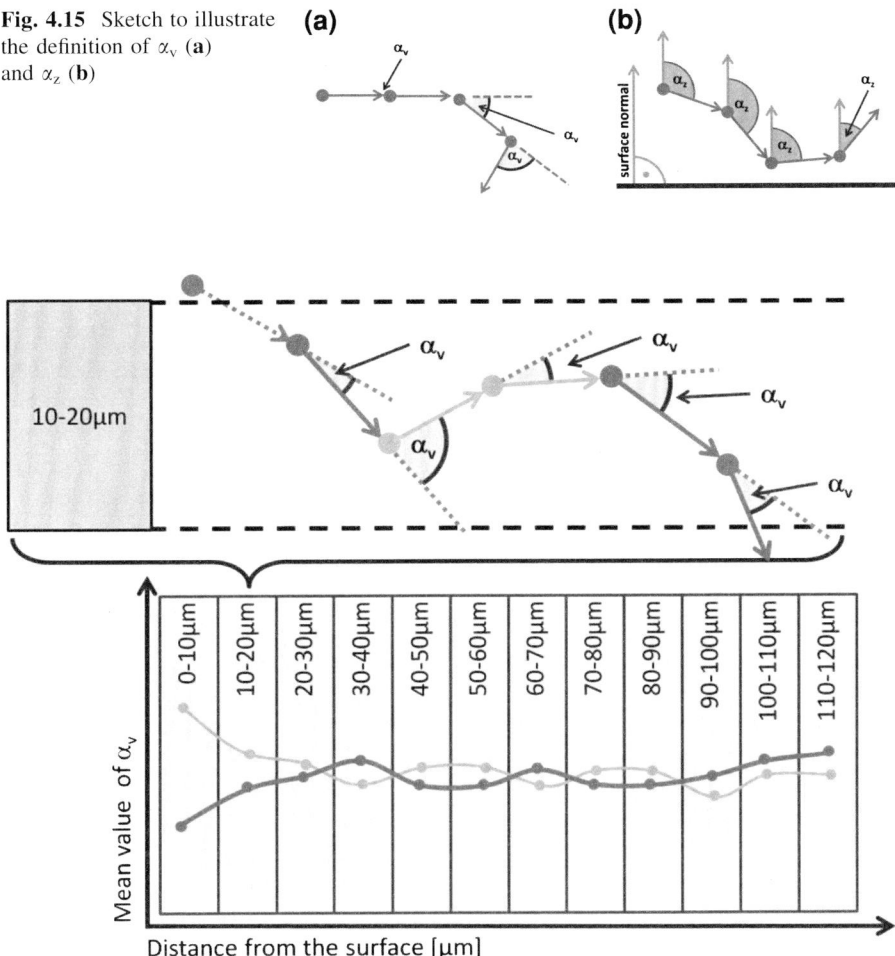

Fig. 4.16 2D sketch to illustrate the motility analysis in dependency to the distance from the surface. The motility vectors marked *dark gray* point towards the surface, whereas the vectors marked *light gray* point away from the surface. Only vectors with a start value within the section are assigned to the section. The start value of the *dotted* vector is outside the section and therefore belongs to the upper section. The average value of all observed vectors is plotted. This sketch is shown as an example for α_v but is done in the course of the thesis for α_z and v_m as well

spore swims away from the surface, and if α_z is bigger than 90° the spore swims towards the surface.

To discuss the spore motility in detail, especially to study the approach and detachment from the surface, the motility parameters (α_v, α_z and v_m) of individual trajectories are determined for different sections of the observation volume. Figure 4.16 shows a theoretical and 2D simplified example for this analysis. The motility vectors are assigned to the different sections dependent on the start point of the vector. The analysis is exemplary shown for the section 10–20 μm from the surface. Only the solid line vectors are assigned to the section. The start value of the dotted vectors is outside the section and therefore does not belong to this section. Based on the value of the α_z, the vectors can be assigned to pointing towards the surface (dark gray) and pointing away from the surface (light gray). The mean value of all vectors in the section is plotted in the corresponding graph (see Fig. 4.16). An example can be found in Fig. 6.3. Respectively, this analysis can be done for α_z and v_m.

Whether a spore swims isotropic in volume can be studied by the means of the mean $\bar{\alpha}_z$ distribution shown in Fig. 4.17. The sketch shows a simplified 2D movement vector distribution. Depending whether the vector points towards the surface the vector is marked dark gray or if pointing away from the surface it is marked light gray. The distribution is isotropic if the average angle is 90° and the average angel in the half spaces is 56° for the spore fraction swimming away and 124° for the spore fraction swimming towards the surface. For the mean angles in

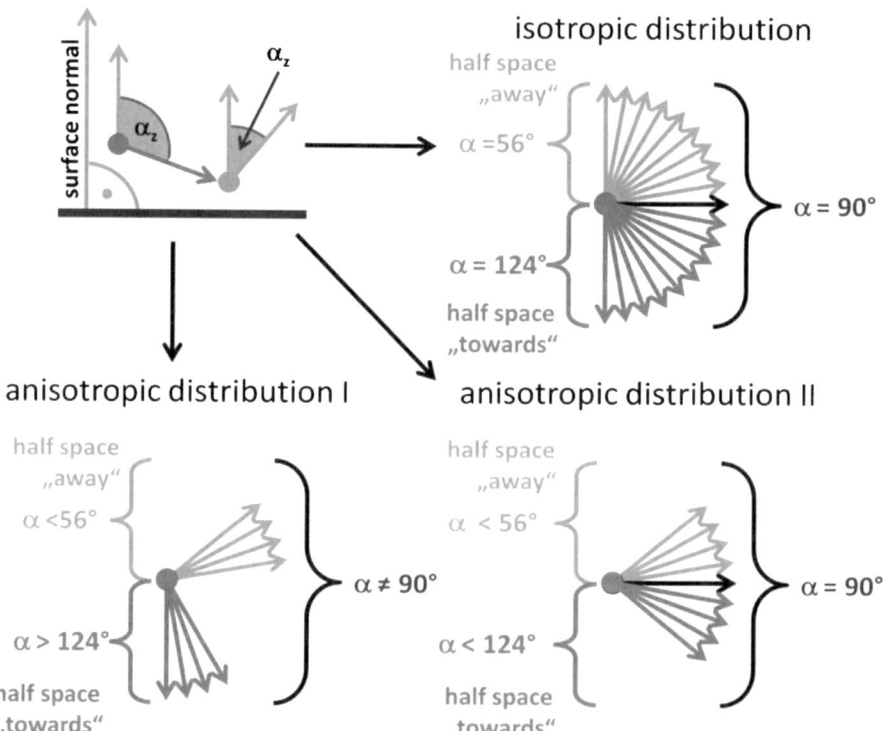

Fig. 4.17 Simplified 2D sketch to illustrate the movement vector distribution. The orientation of the movement vector in correlation to the surface is described by α_z. Depending on the calculation only angles between 0 and 180° are possible. The possible orientation of the vector can be divided in two sections (half spaces). A value of α_z between 0 and 90° means that the spore swims away from the surface (*marked light gray*), whereas a value of α_z between 90 and 180° means that the spore swims towards the surface (*marked dark gray*)

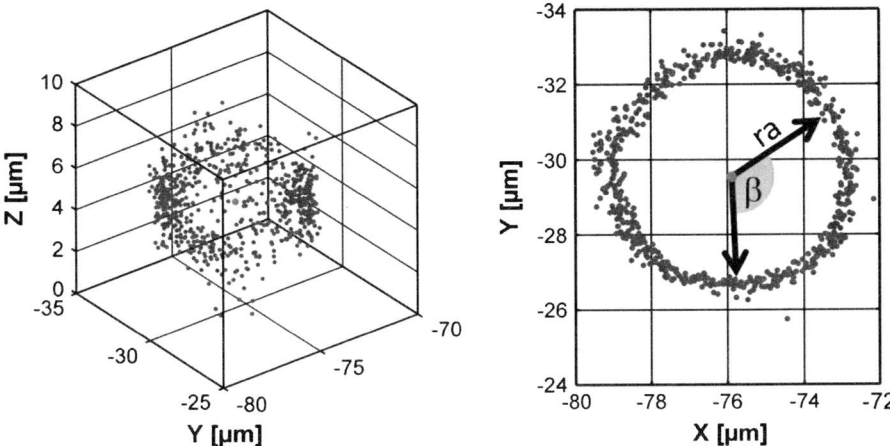

Fig. 4.18 Definition of the radius (ra) and of the angle β to quantify the spinning motion

the half spaces, at a first thought, one would expect a mean value of 45 and 135°, respectively, which is true for two dimensions. However, in a three dimensional volume, it has to be taken into account that a vector has more possibilities of adapting an angle of 90° than of 0°. Therefore, the mean value shifts to 56 and 124°, respectively (see Fig. 4.17, isotropic distribution).

If the mean angle over the complete angle range (Fig. 4.17, anisotropic distribution I) is not 90°, the movement of the spores is anisotropic and a trend to swim either towards or away from the surface is observed. However, if a spore moves parallel to the surface, the mean angle for the complete angle range is still 90° although the spore movement is anisotropic (Fig. 4.17, anisotropic distribution II). This anisotropy can be described by the mean value of $\bar{\alpha}_z$ of the half spaces. If the value is >56° for the movement away from the surface and <124° for the movement towards the surface and the absolute value of difference $|90° - \alpha_{towards/away}|$ is for both angles the same, the spore swims parallel to the surface.

The spinning motion (definition in Sect. 5.2.4.5) is described by the radius (ra) and the angle β. Black dots mark the determined center of mass of the spore bodies. The center of the circle is calculated by the median of the displacement vector. The radius (ra) is the vector which points from the center of the circle to the obtained spore position. The angel β is defined as the angel between two subsequent radial vectors (Fig. 4.18).

4.5 Surface Position Determination

To be able to study the exploration behavior of *Ulva* zoospores on surfaces the exact surface position within the setup (see Fig. 4.1 for a schematic sketch) needs to be known. To determine the surface position is not trivial because the used

surfaces are transparent and are therefore not visible in the reconstruction. One
approach was to scratch the surface on the back side with a diamond pen. Once the
scratch is recorded it is necessary to change the field of view for the motion
analysis because the scratch strongly disturbs the acquired hologram. It is possible
to reconstruct the scratch but the obtained position is not precise enough for the
surface localization. The reason for the insufficient position determination of the
scratch is that it has no defined edges and the light is strongly diffracted. Another
approach was to glue a hair on the backside of the surface. But again the position
determination is not precise enough. Here the reason is that the object is too big to
be recorded with the used magnification. For the reconstruction too much reference wave information is lost to obtain a good image.

As a solution the surface position is determined by obtained motion data itself.
If a spore settles on a surface the surface plane can be easily determined. If no
settlement event is observed during the recording the surface determination is
difficult. Nevertheless, based on the knowledge of the motion data analyzed on
surfaces where settlement occurs it is possible to define the surface plane also on
surfaces on which no settlement is recorded.

Figure 4.19 shows an example of the surface position determination based on
the motility data. With the known uncertainty of the position determination of a
static object (e.g. settlement event, see Fig. 4.14), the surface position can be
found by the calculation of the average z-values of all settlement events. Assuming
the same determination uncertainty the surface position on PEG is defined
respectively.

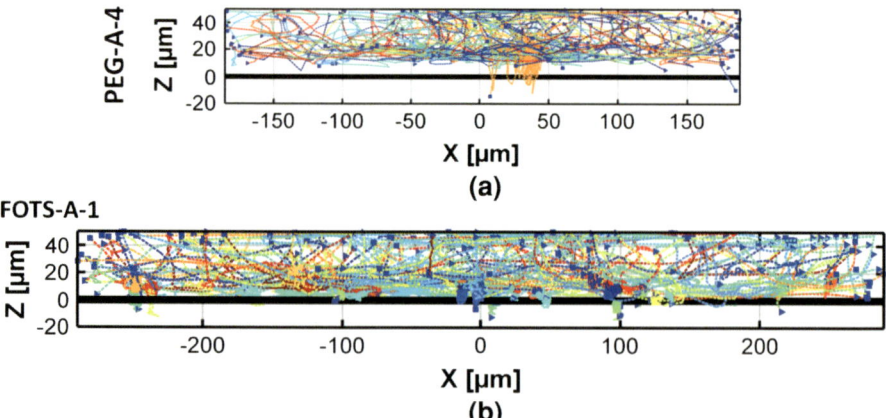

Fig. 4.19 Example to determine the surface position on PEG (**a**) and on FOTS (**b**). The
trajectories are colored to distinguish between the individual trajectories. Please note that the field
of view for FOTS is 1.5 times larger than for PEG. This difference is due to a different pinhole
sample distance

4.6 Experiments with *Ulva* Zoospores

Zoospores where collected and released according to an established protocol [4]. In the thesis of Schilp a detailed description is provided [5]. Fertile plants of *Ulva linza* were collected from the seashore at Llantwit Major, South Wales, UK (51°40′N; 3°48′W) a few days before spring tide. In order to remove any debris which could alter the measurements, the artificial sea water (ASW, Tropic Marin®) and the spore suspension was filtered twice with a 20 μm filter. The optical density of the suspension of released spores was measured and diluted to a final concentration of 20,000 spores/ml. To prevent settlement the spores were kept on a magnetic stirrer for 15 min. This time was held constant to obtain the same conditioned spores for all work phases. The suspension (1 ml) was injected into the dry wet cell and the spore motility was typically recorded for 1 h resulting in 2 Terabyte of storage image per experiment. The recording was started with the injection of the spores and was stopped after the experiment was terminated. The experiment was terminated by sucking the spore suspension out of the wet cell. Afterwards the wet cell was dismounted and the spore settlement on the used surface was counted, or the surfaces were used in a standard settlement assay to determine their performance under established conditions.

The AMBIO standard settlement assay is the following: 10 ml of the spore suspension (1.5×10^6 spores/ml) are added to an individual compartment of a sterile Quadriperm dish containing the test surface of interest. The sample is incubated in darkness for 45 min and then washed gently in ASW to remove unsettled, i.e. motile spores. The spore were fixed in a 2.5% solution of glutaraldehyde in ASW overnight. Subsequently the surface are washed first in ASW, than in 50/50 ASW and deionized water and finally in deionized water. The spores are counted under a Zeiss epifluorescence microscope taking advantage of the autofluorescence of the chlorophyll locate settled spores.

4.7 Investigated Surfaces

Apart from the glass surface all other surfaces were prepared and characterized by X. Cao, a former group member. Within his thesis a detailed description of the preparation and the characterization is provided [6].

In the course of the thesis the following surface are used:

- AWG (glass coverslips: 22×22 mm, 0.14 mm thickness, purchased by Roth GmbH & Co. KG, Germany).
- PEG (polyethylenegycolmonomethylether [MW: 2,000 g/mol with 43 etylenegycol units], synthesized by X. Cao [6]) crafted on glass coverslips.
- FOTS (tridecafluoroctyl-triethoxysilane, purchased from Degussa, Germany) crafted on glass coverslips.

The coverslips were cleaned by rinsing with deionised water, followed by an ultrasonic treatment for 15 min, rinsing with deionised water, leaching for at least 12 h, rinsing with deionised water and finally blow drying with nitrogen. The prepared surfaces (PEG and FOTS) were stored under nitrogen after the preparation and were blown off with nitrogen before the use.

References

1. W. Xu, M.H. Jericho, I.A. Meinertzhagen, H.J. Kreuzer, Proc. Natl. Acad. Sci. USA **98**, 11301–11305 (2001)
2. S.K. Jericho, J. Garcia-Sucerquia, W.B. Xu, M.H. Jericho, H.J. Kreuzer, Rev. Sci. Instrum., **77**(4), 8 pp (2006)
3. M. Heydt, P. Divós, M. Grunze, A. Rosenhahn, Eur. Phys. J. E **30**, 141–148 (2009)
4. M.E. Callow, J.A. Callow, J.D. Pickett-Heaps, R. Wetherbee, J. Phycol. **33**(6), 938–947 (1997)
5. S. Schilp, Self-assembled monolayers and nanostructured surfaces as tools to design antifouling surfaces, Ph.D. Dissertation, Ruprecht-Karls-University of Heidelberg, Heidelberg, 2009
6. X. Cao, Antifouling properties of smooth and structured polyelectrolyte thin films, Ph.D. Dissertation, Ruprecht-Karls-University of Heidelberg, Heidelberg, 2008

Chapter 5
Results: Motility and Exploration Behavior of *Ulva* Zoospores

Understanding the exploration behavior of alga *Ulva* zoospores will give a great insight in how spores select where they settle. Therefore the motility and surface exploration of *Ulva* zoospores is studied. *Ulva* spores are fast swimmers and actively search for a place on a surface to settle for growing into a new plant. The goal of the motion analysis is to understand the surface exploration behavior in deeper detail. This knowledge can be used for example in antifouling assays to distinguish between attractive or repellent surfaces and therefore can assist the development of antifouling coatings. One important condition to further understand the complex surface exploration behavior is to know the motility in solution. Therefore the following chapter is split into two sections: first, the motion analysis in solution and second, the surface exploration behavior.

The experiments with *Ulva* zoospores were done in collaboration with the group of Prof. Callow at the University of Birmingham, UK. Four experimental sessions were carried out in September 2006, April 2007, October 2007 and June 2008, each lasting six weeks. Over this time the setup was refined to fulfill experimental needs and a semi-automatic position determination software package was developed. The data used for the following discussion was acquired in June 2008 because in the earlier recorded data a convection induced flow has been present (see Sect. 4.1.5).

Ulva spores cannot be cultivated and have to be harvested at the sea shore prior to an experiment. The plant only releases a high amount of spores on full or new moon therefore the harvest has to be a few days earlier to obtain sufficient quantities of spores. Data from two collection trips was analyzed.

The following code is used to distinguish between the collection trips. All experiments marked with "***-A-*" were done with spores harvest on June 13, 2008 (5 days before full moon) in Llantwit Major. Experiments marked with "***-B-*" were done with spores collected on June 16, 2008 (2 days before full moon) in Llantwit Major. Table 5.1 shows an overview of the experiment timetable. The spores were released for each experiment separately.

M. Heydt, *How Do Spores Select Where to Settle?*, Springer Theses,
DOI: 10.1007/978-3-642-17217-5_5, © Springer-Verlag Berlin Heidelberg 2011

Table 5.1 Overview over the analyzed experiments

Name	Surface	Collection trip	Day before full moon	Day of the experiment
AWG-I-A-* (Bulk-I-A-*)	AWG	A (June 13, 2008)	5	June 14, 2008
PEG-A-* (Bulk-II-A-*)	PEG	A (June 13, 2008)	5	June 14, 2008
FOTS-A-* (Bulk-III-A-*)	FOTS	A (June 13, 2008)	5	June 13, 2008
AWG-II-B-* (Bulk-IV-B-*)	AWG	B (June 16, 2008)	2	June 17, 2008

The experiments (e.g. "AWG-A-*") are named according to the following system: first, the surface is abbreviated: (e.g.: "AWG": glass, "PEG": poly(ethylene glycol) coated glass and "FOTS": fluorinated monolayer on glass), second, the collection trip is specified (e.g. "A": collection trip A) and third, the experiment number is shown ("*": all experiments, "1": experiment number 1, etc.). For the motility in solution the surface abbreviation is replaced by "Bulk" and a number "I". For all experiments the starting point (t = 0 min) is when the spore suspension is injected into the observation chamber.

For the experiments Bulk-I-A-* and Bulk-II-A-* the spores were released on June 14, 2008 and for the experiment Bulk-III-A-* at the same day as the collection trip (June 13). The spores used for the experiment Bulk-IV-B-* were harvested on June 16, 2008 (collection trip B) and were released at the following day (June 17).

Table 5.2 gives an overview of the analyzed data. In total 662 traces corresponding to 61,146 data points are analyzed. This data is assorted out of 12 individual experiments combining different surfaces and different collections trips. The spore motility is analyzed over an observation time of 11:38 min in total.

5.1 Motility of *Ulva* Zoospores in Solution

To investigate the interactions of spores with a surface it is necessary to understand the motion in solution at first. For this motility study only trajectories are taken into account which are far away from the surface (>200 μm). A detailed discussion about the ability of an *Ulva* spore to sense a surface can be found in Sect. 6.4. These results show that any influence of the surface on the motion can be neglected at distances greater than 200 μm to a surface.

5.1.1 Bulk Motility: Global Analysis of Traces

Figure 5.1 shows some typical swimming patterns of spores in solution. In this figure, 354 individual traces are shown. Bulk-I-A-*, Bulk-II-A-* & Bulk-III-A-* are individual experiments done with spores from the same collection trip. The spores are released for each experiment separately (see Table 5.1 for details).

Table 5.2 Numbers of analyzed traces, data points, and observation time

Name	All		Close to the surface (0–200 μm)		Bulk (200–1,200 μm)		Observation time (s)
	Number of traces	Number of data points	Number of traces	Number of data points	Number of traces	Number of data points	
AWG-I-A-1	40	2,293	21	1,002	25	1,291	41.9
AWG-I-A-2	59	4,206	47	2,523	22	1,683	41.8
AWG-I-A-3	44	3,548	31	2,252	18	1,296	42.0
PEG-A-1	42	2,031	23	774	26	1,257	43.3
PEG-A-2	44	2,335	22	914	26	1,421	41.8
PEG-A-3	33	1,689	18	733	20	956	42.3
PEG-A-4	54	3,768	35	2,110	27	1,658	42.2
FOTS-A-1	49	6,145	17	3,749	44	2,396	55.1
FOTS-A-2	81	14,997	41	10,464	63	4,533	83.8
FOTS-A-3	113	8,779	79	6,099	52	2,680	59.4
AWG-II-B-1	44	4,652	28	3,221	18	1,431	102.8
AWG-II-B-2	57	6,703	51	5,809	14	894	102.4
SUM	660	61,146	414	39,650	354	21,496	698.6

Since the analysis only regards the motility in solution the surface used to seal the wet cell is not important. Nevertheless, the type of surface is listed in Table 5.1.

In Fig. 5.1 the motion data is shown for Bulk-I-A-* (panels (a, e, i)), Bulk-II-A-* (panels (b, f, j)), Bulk-III-A-* (panels (c, g, k)) and Bulk-IV-B-* (panels (d, h, l)). Each experiment is analyzed at various points in time (see Table 5.3) and the complete analyzed motion data for the different points in time (for each experiment) is shown in a single 3D plot. For Bulk-I-A-* (Fig. 5.1, panels (a, e, i)) the total observation time is 2:05.7 min (41.9 s + 41.8 s + 42.0 s). The first trajectories are analyzed after 0:35 min and the last after 11:57 min (see Table 5.3 for details and for the other experiments (Bulk-II-A-*, Bulk-III-A-* and Bulk-IV-B-*)).

All recorded spores shown in Fig. 5.1 move independently from each other in different directions. Some of them move straight, others perform kinks or swim in large circles. No movement into a preferred direction and also, no swarm behavior is detected in the data. Furthermore the data is free of convection.

To obtain a quantitative understanding of the motility data shown in Fig. 5.1, velocity histograms are calculated and displayed in Fig. 5.2. To analyze the time dependency for the experiments one point in time is shown as a histogram for each observation cycle. The relevant values of these time points are summarized in Table 5.4. The vertical lines indicate a velocity of 50, 150 and 250 μm/s and are supposed to help to distinguish changes between the histograms.

For all experiments the velocity distribution changes significantly with elapsing time. For example in Bulk-III-A-1 (panel (g)) analyzed after 0:29 min only one broad peak at a mean velocity of 208 ± 14 μm/s is visible for the velocity

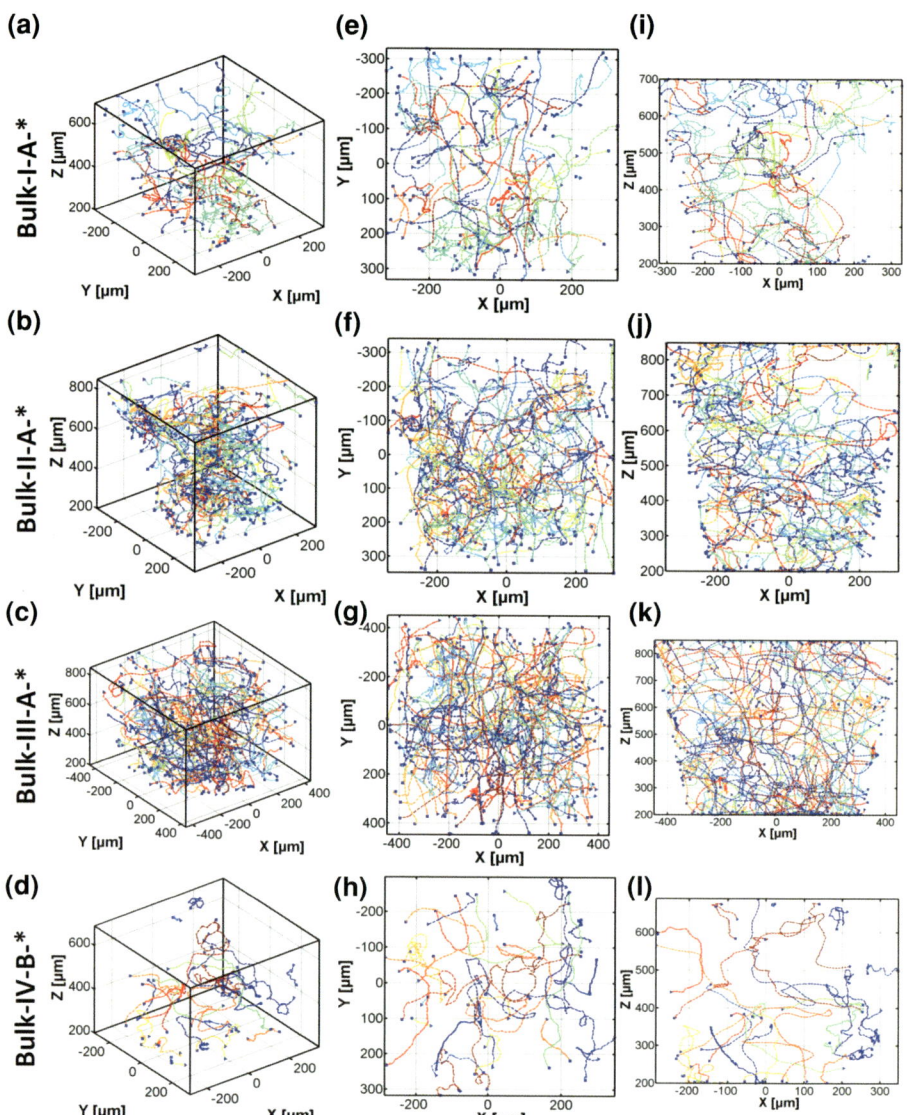

Fig. 5.1 Swimming pattern in solution. Any influence of the surface on the motility can be neglected because only data points which are further than 200 μm from a surface are taken into account. **a–d** 3D view; **e–h** xy view; **i–l** xz view. The trajectories are colored differently for a better differentiation between individuals trajectories

distribution. 0:55 min later in Bulk-III-A-2 (panel (h)) two clearly distinguishable peaks are present in the histogram. The position of peak (I) is 57 ± 7 μm/s and of peak (II) 224 ± 14 μm/s. The position of peak (II) is within the error unchanged compared to the position observed in Bulk-III-A-1 whereas the slower peak

Table 5.3 Details for the experiments in solution

Name	Elapsing time (min)	Observation time (s)
Bulk-I-A-1	0:35	41.9
Bulk-I-A-2	3:14	41.8
Bulk-I-A-3	11:57	42.0
Bulk-II-A-1	1:26	43.3
Bulk-II-A-2	2:09	41.8
Bulk-II-A-3	2:51	42.3
Bulk-II-A-4	6:54	42.2
Bulk-III-A-1	0:29	55.1
Bulk-III-A-2	1:24	83.8
Bulk-III-A-3	6:24	59.4
Bulk-IV-B-1	5:00	102.8
Bulk-IV-B-2	22:39	102.3

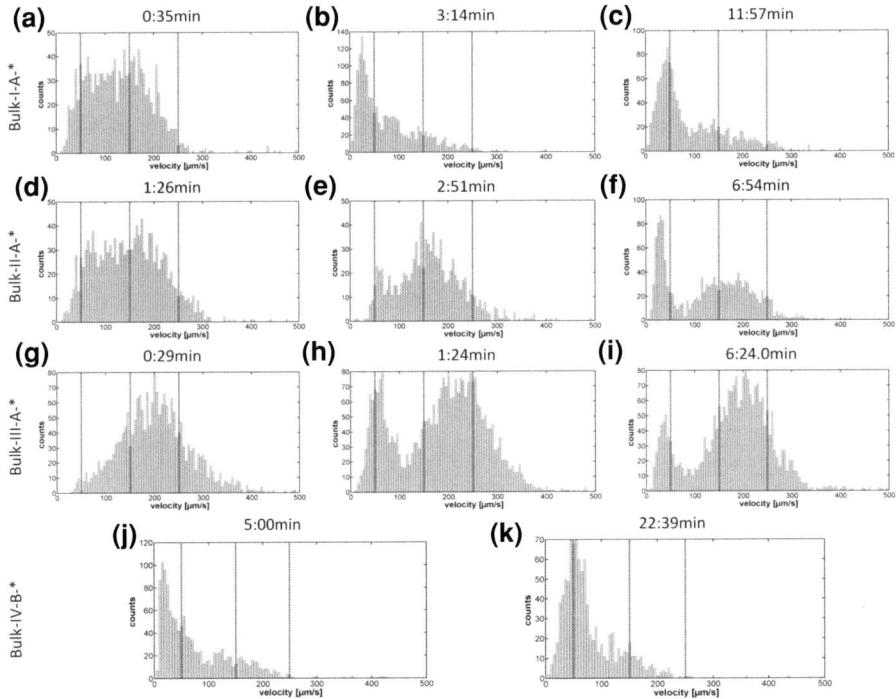

Fig. 5.2 Velocity histograms for the analyzed spores in solution. **a** Bulk-I-A-1 (0:35 min); **b** Bulk-I-A-2 (3:14 min); **c** Bulk-I-A-3 (11:57); **d** Bulk-II-A-1 (1:26 min); **e** Bulk-II-A-3 (2:51 min); **f** Bulk-II-A-4 (6:54 min); **g** Bulk-III-A-1 (0:29 min); **h** Bulk-III-A-2 (1:24 min); **i** Bulk-III-A-3 (6:24 min); **j** Bulk -IV-B-1 (5:00 min); **k** Bulk-IV-B-2 (22:39 min). The *vertical lines* indicate velocities of 50, 150 and 250 μm/s

Table 5.4 Details for the histograms shown in Fig. 5.2 (RSF: Ratio between slow and fast spores)

Experiment	Panel	Elapsing time (min)	Peak (I) (µm/s)	Mean peak (I)	Peak (II) (µm/s)	Mean peak (II)	RSF
Bulk-I-A-1	a	0:35	51 ± 7		156 ± 12		0.4
Bulk-I-A-2	b	3:14	36 ± 6		149 ± 11		1.8
Bulk-I-A-3	c	11:57	43 ± 6	44 ± 8	167 ± 12	157 ± 9	1.6
Bulk-II-A-1	d	1:26	57 ± 7		172 ± 13		0.3
Bulk-II-A-3	e	2:51	30 ± 5		174 ± 12		0.4
Bulk-II-A-4	f	6:54	30 ± 5	42 ± 15	173 ± 12	170 ± 6	0.4
Bulk-III-A-1	g	0:29	–		208 ± 14		0
Bulk-III-A-2	h	1:24	57 ± 7		224 ± 15		0.4
Bulk-III-A-3	i	6:24	42 ± 6	49 ± 10	205 ± 14	212 ± 10	0.3
Bulk-IV-B-1	j	5:00	35 ± 6		154 ± 11		2.1
Bulk-IV-B-2	k	22:39	55 ± 7	45 ± 14	158 ± 10	156 ± 7	4.0

appears with increasing time. For Bulk-III-A-3 (panel (i)) after 5:54 min the two peaks are also clearly distinguishable. With respect to the already listed peak positions for Bulk-III-A-1 and Bulk-III-A-2, the position of peak (II) (205 ± 14 µm/s) is unchanged whereas the position of peak (I) is shifted to lower velocities (42 ± 6 µm/s). The same trend—two clearly distinguishable peaks developing with increasing time—holds true for the other three experiments, too. In Table 5.4 the peak positions for each histogram are shown.

Interestingly, the value for peak (I) which is listed in Table 5.4 is fairly constant for all experiments, whereas peak (II) is only constant within an experiment. Also the ratio between the amount of slow and fast spores (RSF) is different for the individual experiments and changes with increasing observation time (see Fig. 5.3). For this analysis, a threshold of 100 µm/s was used to differentiate fast and slow spores. The threshold was chosen such that it corresponds to the average of the minima in the velocity histograms shown in Fig. 5.2. For example, in Bulk-IV-B-* (red curve), after 300 s, twice as many data points are assigned to slow spores as assigned to fast spores. For Bulk-IV-B-* with increasing experimental time the ratio increases. After 1,300 s the RSF rose to four times more slow spores than fast spores. In general the amount of slow spores in the bulk increases with elapsing time (see Fig. 5.3, panel (a)).

Furthermore the RSF is related to the mean velocity (v_m) of the fast spore fraction. The faster the mean velocity (peak II) the smaller the RSF (see Table 5.4 or Fig. 5.3, panels (a, b)). However, the v_m depends on the time of release of the spores (see Fig. 5.3, panel (c)). The earlier the spores are released after the harvest at the sea shore and the shorter the collected leaves are stored in a fridge, the faster the released spores are able to swim. Furthermore the RSF can also be related to the storage time of the leaves after the harvest. No significant divergence is observed for the mean velocity of the fast spore fraction released from leaves harvested at different collection trips (see Fig. 5.3, panel (b)).

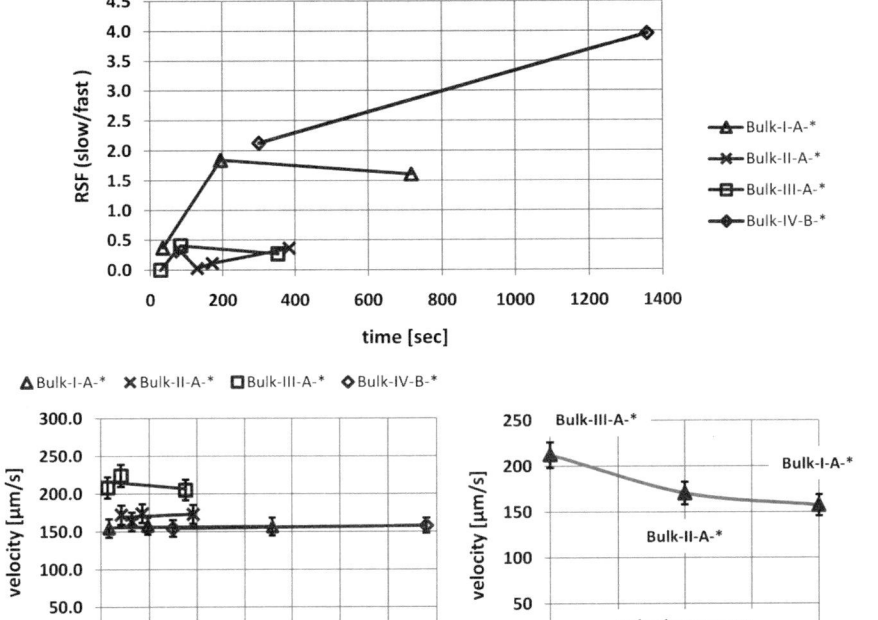

Fig. 5.3 Comparison of the spore performance between the individual experiments: **a** number of slow (v < 100 μm/s) spores divided by number of fast (v > 100 μm/s) spores (RSF); **b** comparison of v_m for the fast spores harvested at different collection trips (A and B). The exact release time after the collection trip is summarized in Table 5.1. For collection trip A the mean value of v_p for Bulk-I-A-*, Bulk-II-A-* and Bulk-III-A-* is used; **c** v_m for the fast spores harvested at collection trip A versus storage time

Figure 5.4 shows a color coded version of Fig. 5.1 to distinguish between slow (marked red) and fast (marked blue) spores. Interestingly, each trace of a spore is either slow or fast. In the complete data set not a single trace is detected which shows a switch between slow and fast fraction.

5.1.2 Bulk Motility: Detailed Motion Analysis for Individual Traces

Of all traces discussed in the section above, five representative traces are chosen for detailed motion analysis. The discussed traces are chosen from different experiments to demonstrate that the trends are identical. The analysis is mainly based on the 3D rendered plot, velocity histogram and a detailed angle analysis. Each velocity histogram is fitted with a Maxwell–Boltzmann distribution (g(x) Eq. 5.1) [1] with the free parameters a, b and c:

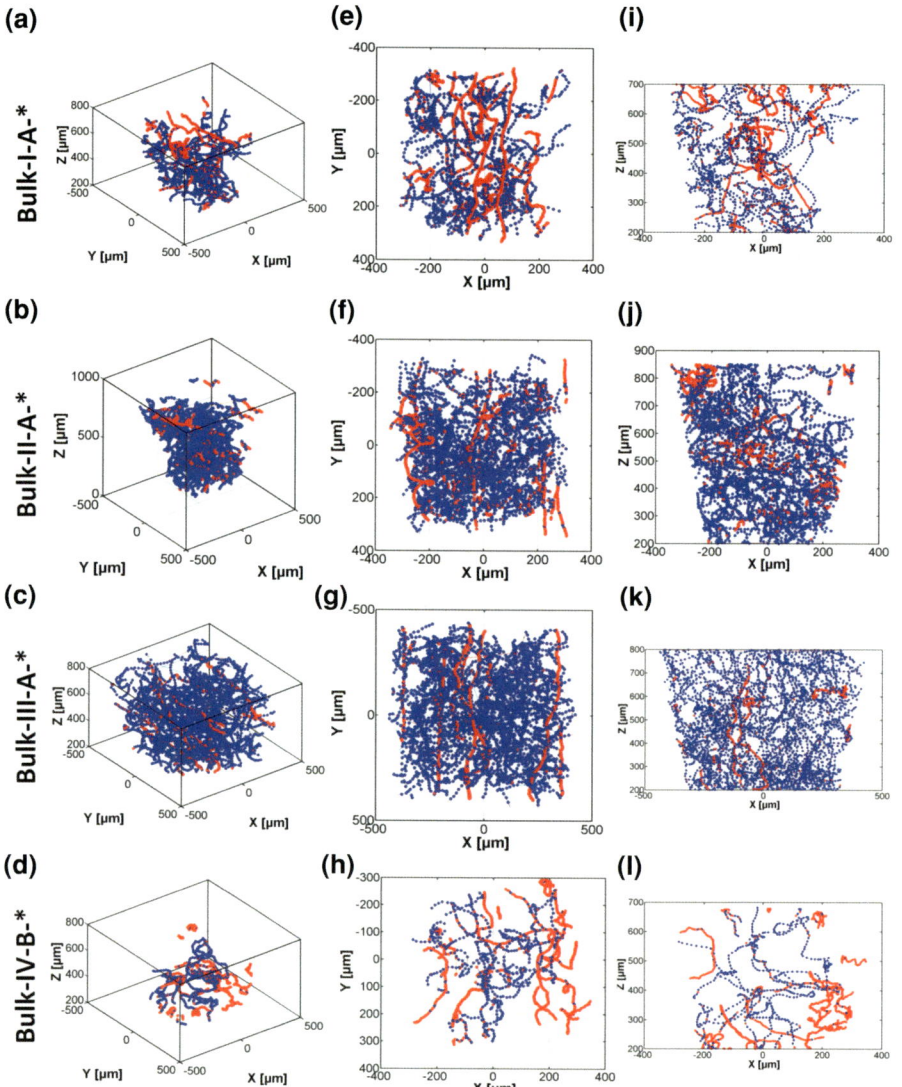

Fig. 5.4 Color coded swimming pattern in solution (•: v < 100 μm/s; •: 100 μm/s < v < 500 μm/s); **a–d** 3D view; **e–h** xy view; **f–l** xz view

$$g(x) = ax^2 e^{-\left(\frac{x-b}{c}\right)^2} \tag{5.1}$$

The Maxwell–Boltzmann distribution is derived from the Boltzmann distribution of energies and is valid for an ideal gas [1]. It describes speeds of gases, where the particles do not interact with each other aside from collisions. The objects move freely between the collisions. In the experiment described here the spore concentration is very low and according to Fig. 5.2 it the spores move independent of each other. Therefore the Maxwell–Boltzmann can be used and it describes the *Ulva* spore velocity distribution very well.

This section is split into two subsections. In the first part, the spore fraction which moves fast is discussed, the second part deals with the slower spore fraction.

5.1.2.1 Fast Spore Fraction

In Fig. 5.5, five individual traces are shown exemplarily. The trajectories are named according to the following system: "Bulk-III" represents the experiment (as described in the previous section) and "FaS-xy" denotes: **Fa**st **S**pore number **xy**. Figure 5.5 shows that all spores steadily move forward and do not stay at certain positions. The corresponding velocity histograms are shown in Fig. 5.6, panels (f–j). For the five exemplary traces the most probable velocity (v_p) and the full width at half-maximum (FWHM) are summarized in detail in Table 5.5.

As already described in Sect. 5.1.1, for the general velocity histograms there is a significant divergence for v_p between the experiments. But even for an individual experiment the difference between single spores can be large (see Table 5.5, Bulk-III-FaS-1: 202 µm/s, Bulk-III-FaS-3: 253 µm/s). For a fast spore the FWHM is $44 \pm 12\%$ of v_p. Even though this number is high for an individual trajectory, no longer fast and slow movement phases are observed. In the motion pattern no run or tumble phases as in the movement of for example *E. coli* (see Sect. 3.3.2 for details about run and tumble) are found. As explained in Sect. 5.1.1 (see Fig. 5.3, panel (b)) the mean velocity for the fast spore fraction (within an individual experiment) does not change with increasing recording time, only the ratio between slow and fast spores is increasing with elapsing time. For the individual traces it is not observed that a fast swimming spore is getting "tired" (slower) while it swims in solution (see histogram in Fig. 5.2 or Fig. 5.3, panel (b)). The different values in Table 5.5 for v_p show that individual fast spores have different capabilities how fast they can swim. Some are able to swim very fast (e.g. Bulk-III-FaS-3: 254 µm/s) and others are slower (e.g. Bulk-III-FaS-1: 203 µm/s).

Figure 5.7 shows the swimming pattern of spores assigned to the fast spore fraction in the bulk. In this figure, the spore position marker is colored according to the velocity. The color gradient encoding different velocities spans from 50 µm/s (light blue) up to 500 µm/s (pink). Based on the histograms shown in Fig. 5.1 50 and 500 µm/s are used as a threshold to describe the velocity distribution. For most trajectories shown in this figure it is observed that the spores swim slower when they perform a turn. The slowdown is apparent in a blue shift of the individual trajectory. The observation that spores swim slower when performing a turn can explain the broad velocity histogram of the individual spores (see Fig. 5.6).

Furthermore, to describe the swimming characteristics of the spores the angular distribution of α_v is plotted in Fig. 5.6, panels (k–o). The angle α_v (see Fig. 4.15) is the angle between two consecutive velocity vectors. If for example α_v is $0°$ the spore swims in a straight line and if α_v is $180°$ it swims backwards. In Fig. 5.6,

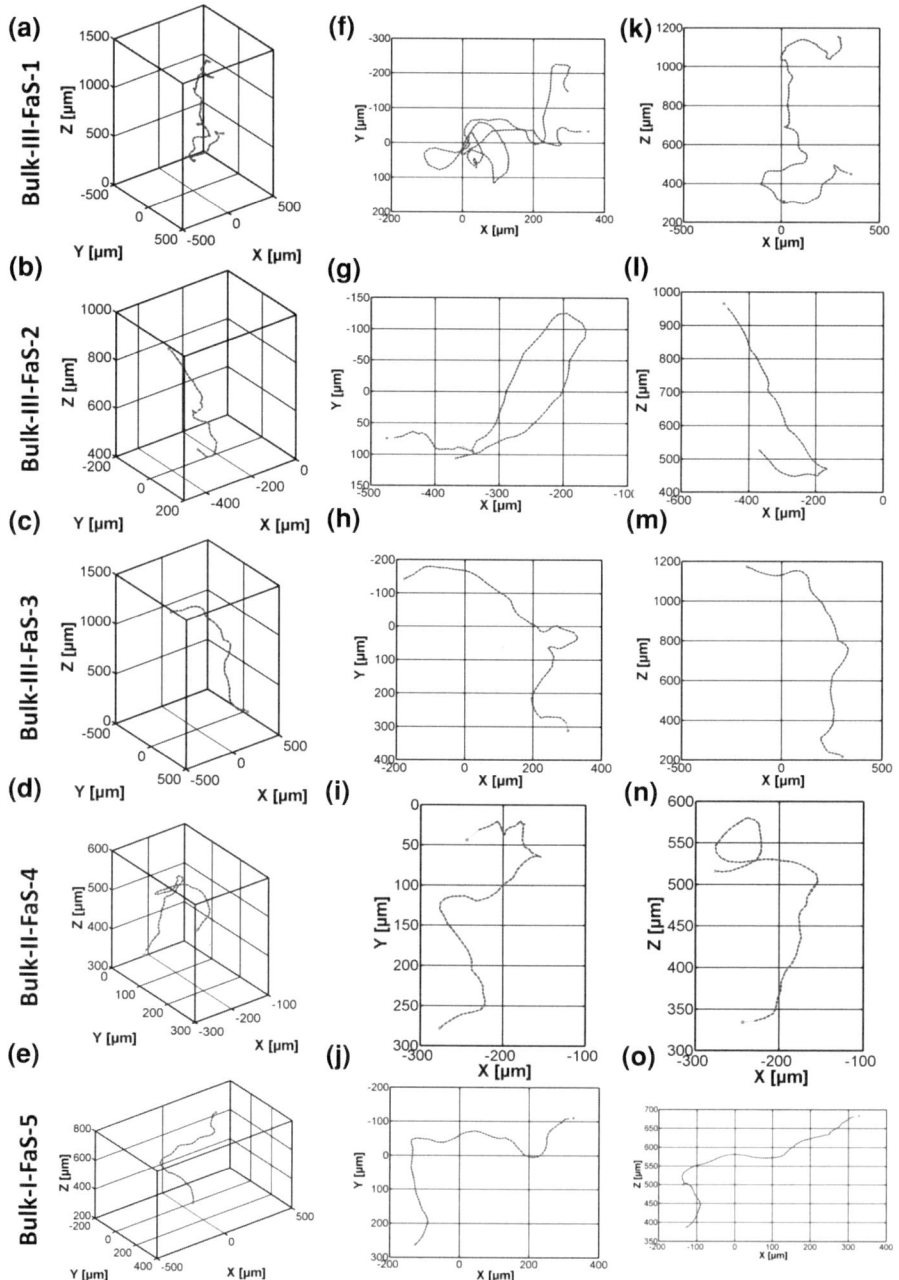

Fig. 5.5 Five fast swimming individual and exemplary traces: **a–e** 3D view; **f–j** xy view; **k–o** xz view

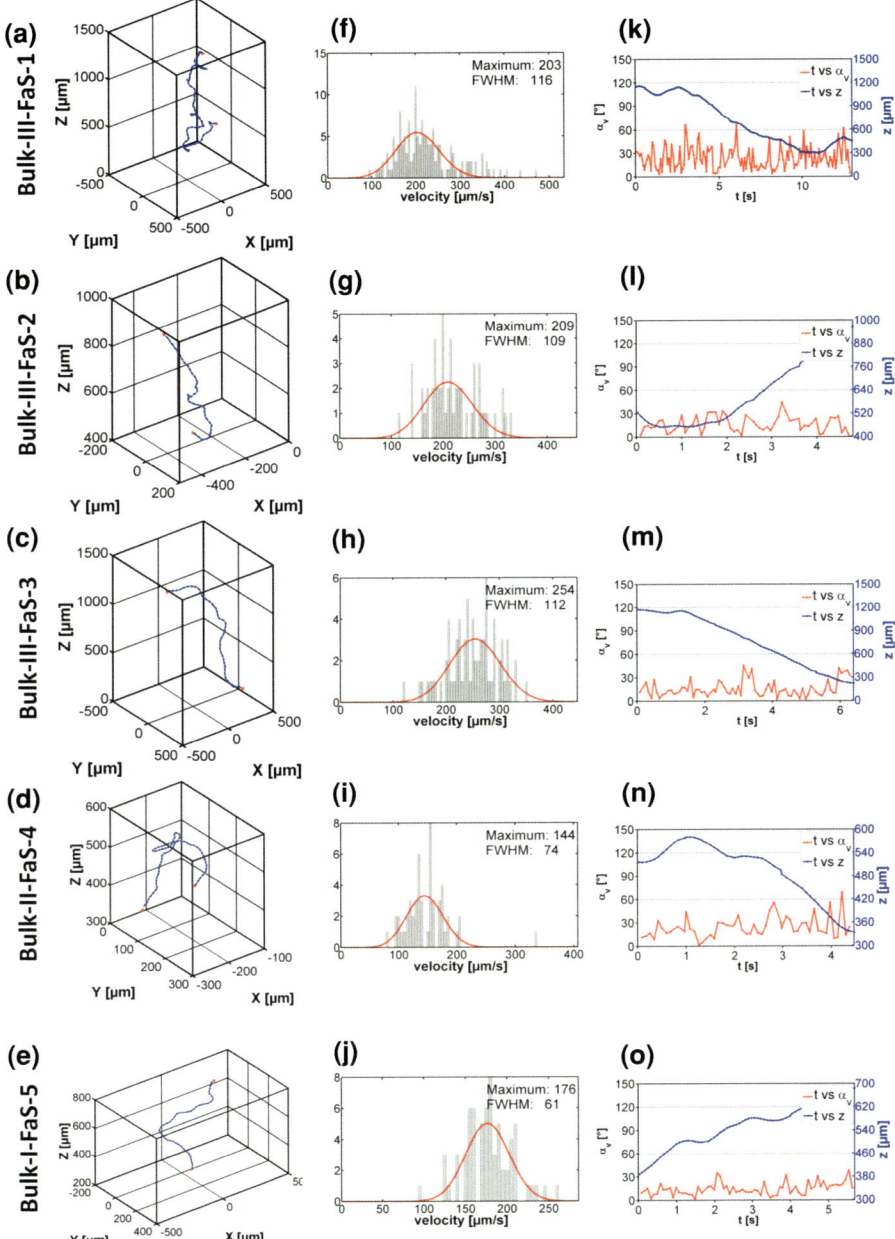

Fig. 5.6 Five fast swimming individual and exemplary trajectories: **a–e** 3D view; **f–j** velocity histogram of the trajectory with fitted Maxwell–Boltzmann distribution; **k–o** elapsed time versus angle distribution (α_v, *red*, scale on *left side*) and distance to the surface (*blue*, scale on *right side*). The meaning of α_v is explained in Fig. 4.15

Table 5.5 Comparison of the individual fast spores ($v_p \pm$ half width at half-maximum (HWHM)) to v_m of the fast spore fraction of the corresponding experiment (see Table 5.4)

vp (µm/s)	Bulk-III-FaS-1	Bulk-III-FaS-2	Bulk-III-FaS-3	Bulk-II-FaS-4	Bulk-I-FaS-5
Single trace	203 ± 58	209 ± 55	254 ± 56	144 ± 37	176 ± 30
Complete experiment	212 ± 14	212 ± 14	212 ± 14	170 ± 12	157 ± 12

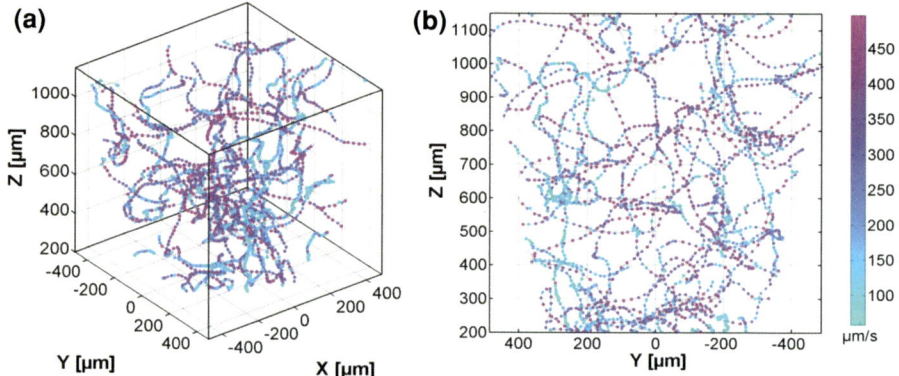

Fig. 5.7 Motility of fast spores in solution color coded for different velocities. **a** 3D view and panel; **b** xz view

panels (k–o) α_v (red, scale on left side) and the z distance to the surface (blue, scale on right side) is plotted against the observation time. For the traces Bulk-III-FaS-2, Bulk-III-FaS-3, Bulk-II-FaS-4 and Bulk-I-FaS-5 α_v it is $\approx 20°$ with only a few values higher than $40°$ (see Fig. 5.6, panels (l–o)). For Bulk-IIII-FaS-1 the values of α_v are slightly higher but none of them is larger than $70°$ (see Fig. 5.6, panel (k)). For a fast spore α_v is typically in the range from $5°$ to $60°$ which means that the spore does not move in a straight line but with a strong forward preference. The change in z-position as shown in Fig. 5.6 (blue curve) is not so relevant for the discussion in the bulk but will get more important in surface exploration Sects. 5.2 and A.1–A.3.

5.1.2.2 Slow Spore Fraction

To characterize the slow spore fraction five exemplary traces are shown in Figs. 5.8 and 5.9. The trajectories are named following the same systematic used for the fast spores. The name consists of the name of the experiment (e.g. Bulk-I) and an abbreviation for **S**low **S**pore number **xy** (SlS-xy). In Fig. 5.8 the motion patterns in solution are shown. The 3D rendered plots appear similar to the motion pattern shown in Fig. 5.5 for the fast spores. Slow spores, in comparison to fast spores, swim significantly slower (see Fig. 5.9, panels (f–j)) and fidgety around a position. The most probable velocities (v_p) are summarized in Table 5.6.

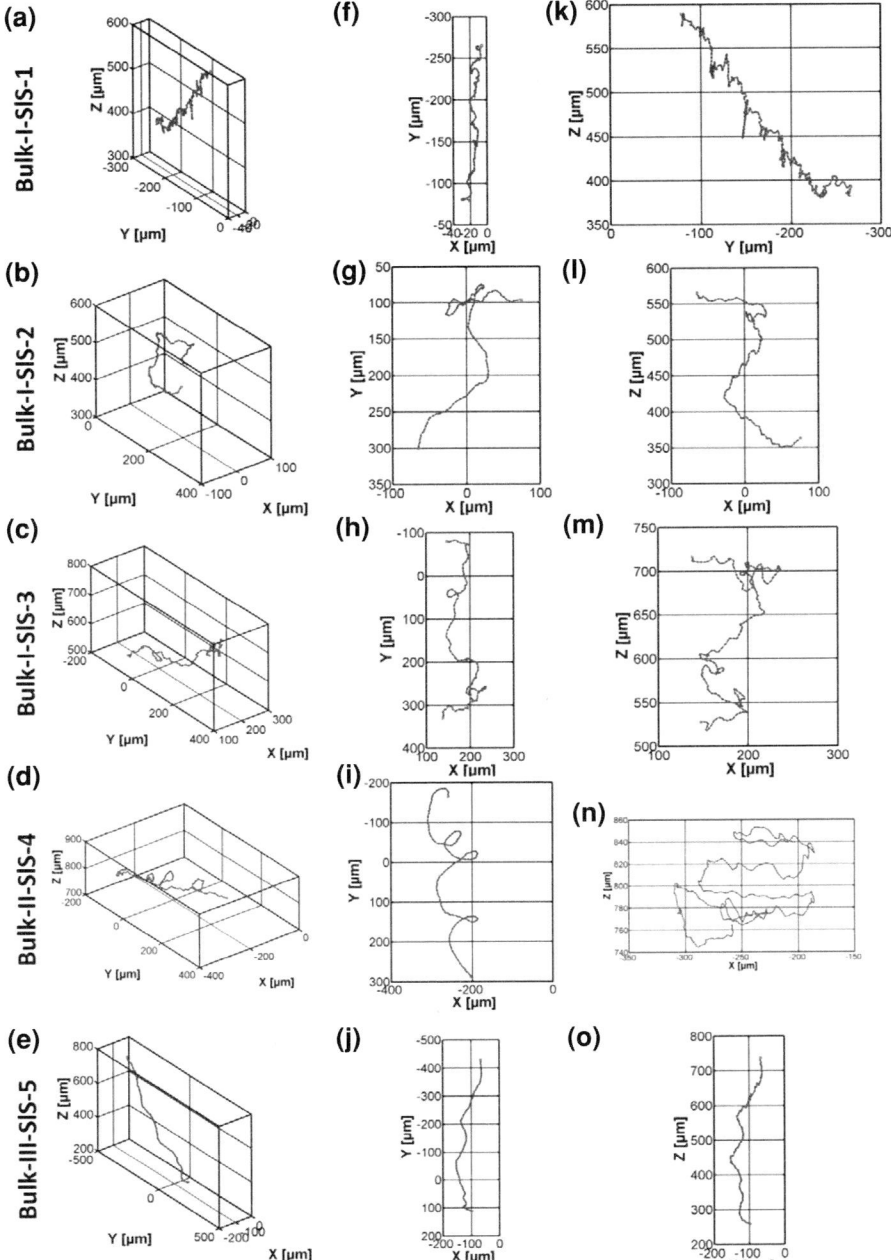

Fig. 5.8 Five slow swimming individual and exemplary traces: **a–e** 3D view; **f–j** xy view; **k–o** xz view

Table 5.6 Comparison of the $v_p \pm$ HWHM for the individual slow spores and v_m of the slow spore fraction of the corresponding experiment (see Table 5.4)

v_p (µm/s)	Bulk-I-SlS-1	Bulk-I-SlS-2	Bulk-I-SlS-3	Bulk-II-SlS-4	Bulk-III-SlS-5
Single trace	19 ± 15	29 ± 12	52 ± 27	30 ± 13	35 ± 18
Complete experiment	44 ± 8	44 ± 8	44 ± 8	42 ± 15	49 ± 10

Between the spores which are assigned to the slow spore fraction a significant divergence for v_p is observed. This observation is exemplary explained for traces analyzed within experiment Bulk-I (see Table 5.6). The trace Bulk-I-SlS-1 is slow ($v_p = 19 \pm 15$) whereas the trace Bulk-I-SlS-3 is quite fast ($v_p = 52 \pm 27$) for a slow trace. For the other experiments (Bulk-II, Bulk-III and Bulk-IV) differences between the velocities of individual traces are also observed.

The slow spore fraction does not only differ from the fast fraction in speed but also concerning the α_v distribution. In Fig. 5.9, panels (k–o) the distribution of α_v (red, left side) and the change in z-position (blue, right side) are plotted against the observation time. The obtained values for α_v range between $10°$ and $170°$ for all analyzed spores. This means that a spore does not swim forward (like it is observed for a fast moving spore) but rather in an erratic motion.

5.1.3 Summary of the Motility in Solution

From the observed data the following conclusions can be drawn:

- There are two distinguishable spore fractions (fast and slow) in solution.
- Spores have different capability how fast they can swim.
- No swarm behavior was found in the motion data. The spores move independently from each other.
- While a spore performs a turn it is slower than swimming straight.
- Within the recorded trajectories spores do not switch between the fast and the slow spore fraction.
- Fast spores swim with a strong forward preference ($\bar{\alpha}_v = 28 \pm 25°$).
- Slow spores swim also forward but do change their direction of motion with a high frequency ($\bar{\alpha}_v = 38 \pm 30°$).

5.1.4 Discussion of the Motility in Solution

The 3D motion analysis of the swimming behavior of *Ulva* zoospores in solution shows that two different kinds of swimming patterns (slow and fast spore fraction) are observed. Several potential explanations are discussed to understand the occurrence of the different kinds of spore fractions.

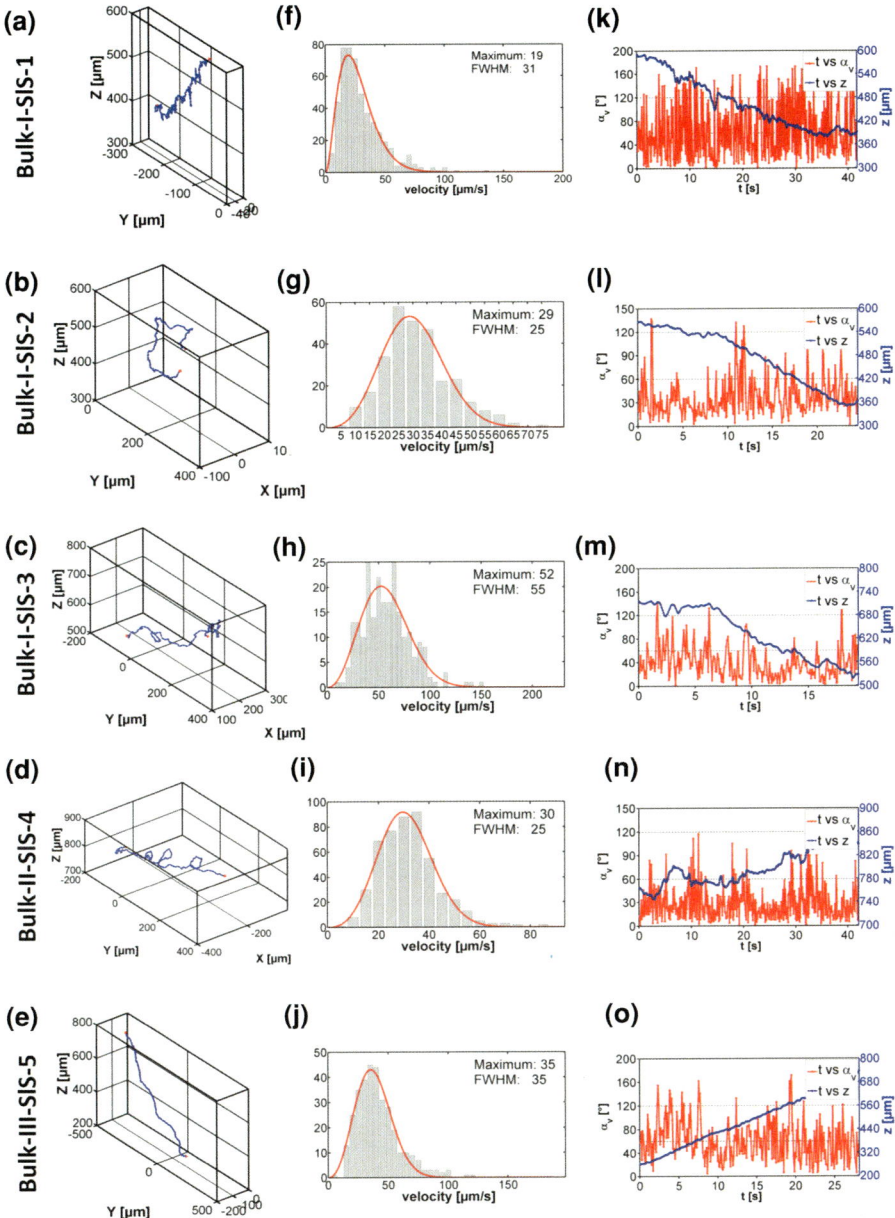

Fig. 5.9 Five slow swimming individual and exemplary trajectories: **a–e** 3D view; **f–j** velocity histogram with fitted Maxwell–Boltzmann distribution; **k–o** time versus angle distribution (*red, left side*) and distance to the surface (*blue, right side*)

In general, the ratio between slow and fast spores (RSF) and the v_p of the fast spore fraction is different for each release and correlates with the time passed after the spore harvest. The sooner the spores are used, the faster they are able to swim and the less slow spores are observed. The difference in the performance between individual releases is on the same order as between different collection trips. The data show that individual spores swim with a constant speed that differs from spore to spore.

One potential explanation for the different spore fractions is that the spores are able to swim in different modes. This is supported by the motility studies of Chlamydomonas [2–4]. After photoshock, *Chlamydomonas* swim in a straight line backwards. This reverse movement only lasts for a short period and ends typically with a randomization of the direction of movement and subsequent with a normal forward swimming (for more details see Sect. 3.3.2). This kind of motion is not observed for the slow spore fraction, because a typical slow spore swims according to the "slow" motion pattern for a long time, meaning that it does not swim in a straight line but rather in a changeful swimming motion. For *Cymbomonas tet-ramitiformis* it is observed that the alga is able to swim in two different modes which occur with a similar probability [5]. *Cymbomonas t.* changes frequently between the different swimming modes. However, no switch between fast and slow swimming spore fraction is observed within the motility study of *Ulva*, it could still be possible that spores are able to swim in two modes. The reason that no switch is observed could be that the switch is irreversible and therefore the probability for the observation is very low.

The slow mode swimming might be a more passive motion in which the spore saves energy, or does not have enough energy for the active exploration. This hypothesis is supported by the fact that spores have limited capabilities to produce energy. When they are released from the plant they have a certain amount of energy to find a place to settle and to grow to a new plant. If this energy is consumed they die.

Another theory is that the slow spores are not spores but rather gametes (see Sect. 3.1 for details of the *Ulva* live cycle). Gametes only have two flagella and therefore might be not able to swim as fast as spores with four flagella. Gametes are positive phototactic and swim actively towards light, whereas spores are negative phototactic. Spores and gametes are released at the same time. The two types can be separated because spores accumulate at the bottom of the flask whereas gametes tend to swim towards the water air interface. The fact that the amount of slow spores increases with elapsing experimental time can be understood if the slow spores are gametes. The field of view (FoV) is the brightest location in the wet cell because it is illuminated by the laser for the recording. Gametes would swim into the field of view within elapsing time because they are positive phototactic and accumulate whereas the spores tend to move away from the light out of the FoV. Thus the RSF increases.

A fourth explanation is that some spores are probably damaged by the stirrer, or during the release from the plant, or even at their "production" and therefore are slower. In the study of *H. irregularis* (a brown algae) by Iken et al. [6] a slow spore fraction is also observed. In this study the slow spore fraction is most probably explained with damaged spores. In Sects. A.1–A.3 another explanation why the RSF in solution increases with elapsing time is shown. The motility data can be interpreted in the way that the fast spore fraction accumulates near any surface whereas the slow spores stay in the water column. Therefore the slow spore fraction remains in solution and increases in concentration.

Concluding, the origin of the slow spore fraction cannot be unambiguously identified. At the moment we favor the last explanation.

5.2 Surface Exploration

The result of fouling is easily recognizable on a macroscopic level, but the effects that lead to it, surface location, exploration and adhesion occur for *Ulva* spores on a micrometer length scale. To study the exploration behavior and the settlement process in situ, three surfaces with different attractiveness for spores are chosen. The used surfaces are already described in detail in Sect. 4.7. However a short summary is given in the following section.

5.2.1 Standard Settlement Study

The settlement of spores on the used surfaces is determined by a standard settlement assay which is used in the AMBIO project to evaluate potential antifouling coatings [7]. The assay is in detail described in Sect. 4.6. In briefly, 10 ml (1.5×10^6 spores per ml) of freshly released spores are incubated in darkness for 45 min. Subsequently the surfaces are washed gently to remove unsettled spores. After fixation and drying, the spores are counted under an epifluorescence microscope to determine the total settlement. The result of such a settlement assay for the three surfaces is shown in Fig. 5.10.

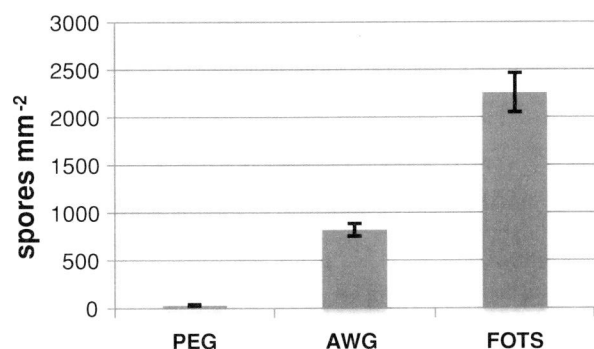

Fig. 5.10 Spore settlement per mm^2 on the investigated surfaces (*PEG* poly(ethylene glycol) coated glass surface; *AWG* glass; *FOTS* tridecafluoroctyl-triethoxysilane monolayer on glass)

The glass surface (AWG) was selected because it is widely used as a standard surface to compare settlement and adhesion studies. With respect to the settlement on the other investigated surfaces AWG is an intermediately attractive surface. The second surface is a poly(ethylene glycol) (PEG) coated surface which is protein resistant and anti adhesive for *Ulva* zoospores—as long as the surface is stable (≈ 13 h) [8]. The third surface is a fluorinated self assembled monolayer on glass (FOTS) where the overall fouling is fairly high, but—which makes the fluorinated coatings commercially interesting—the adhesion strength of an organism is very low (fouling release coating) [9].

5.2.2 Expected Spore Settlement During a Holographic Tracking Experiment

The standard settlement assays are used to determine the time dependence of the settlement on the investigated surfaces. The settlement is determined after 10, 20, 30 and 60 min and the amount of settled spores is counted at 30 positions on each slide to obtain the average settlement for the complete surface. Figure 5.11 shows the results of this study. The study was done within the Diploma thesis of Isabel Thomé in our group [10]. The FOTS and PEG chemistries are grafted onto a gold substrate by a thiol linkage whereas the surfaces used for the holographic tracking are grafted on glass via a silane coupling to obtain a better transparency of the surfaces.

The settlement increases approximately linearly with elapsing time. The slope for the increase is significant different for the different surfaces. With this knowledge the expected settlement for the holographic experiment can be calculated on the basis of the assumption that the settlement is homogeneous over the complete surface. This assumption is important because for holographic recording the field of view (FoV) is fixed and has a size of only a few square hundred micrometers (160–340 μm^2). Holographic recording is only feasible with a much lower spore concentration (max 2×10^4 spores/ml) than used in the standard assay (1.5×10^6 spores/ml) [11]. Based on the approximated linearity of the curves we assume that the spores settle with the same rate even at these low concentrations. The duration of a typical holographic experiment is max 40 min. In this time period the increase in settlement is linear with the elapsing time (see Fig. 5.11).

$$S_{(t)} = \text{FoV} \cdot r \cdot t \frac{C}{C_s} \qquad (5.2)$$

The expected settlement for the concentration used in holography is summarized in Table 5.7 and calculated according to Eq. 5.2 where $S_{(t)}$: settlement [spores]; FoV: field of view [mm^2]; r: linear increase observed for the standard settlement assay with the standard assay concentration $\left[\frac{\text{spores}}{mm^2\,\text{min}}\right]$; t: time [min];

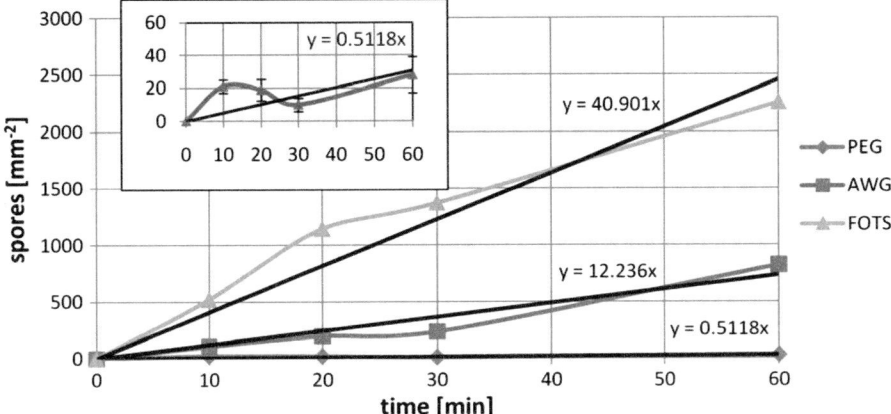

Fig. 5.11 Time dependent settlement analysis for FOTS, AWG and PEG (see inset) surface. The settlement increases linear with elapsing time [10] (black lines linear fit)

	Time (Min)	PEG	AWG	FOTS
Table 5.7 Number of expected settlement events during the holographic recording for the investigated surfaces and the imaged field of view (FoV)	1	0.0	0.0	0.2
	5	0.0	0.1	0.9
	10	0.0	0.3	1.9
	20	0.0	0.5	3.7
	40	0.0	1.0	7.4

C: holographic recording concentration $\left[\frac{\text{spores}}{\text{ml}}\right]$; C_s: standard assay concentration $\left[\frac{\text{spores}}{\text{ml}}\right]$.

The values shown in Table 5.7 illustrate that it is difficult to monitor a settlement event on the resistant surface PEG and also on AWG. This is due to the low concentration and the small FoV in the experiment. However, it was possible to record settlement events at the FOTS and even on the AWG surface.

5.2.3 Settlement Analysis on the Investigated Surfaces

It is important to know whether settlement occurs on the investigated surfaces or not. During the measurement it is possible to witness the spore motility directly in the holographic diffraction pattern. During the lifecycle of an *Ulva*, spore settlement is a crucial step and if settlement occurs—during the recording—it is possible to make the assumption that the used spores are "healthy" and explore the surface as they do in the standard lab bench assays. The settlement during the holographic recording is determined by a direct analysis of the holograms without a reconstruction. The following analysis is demonstrated only for the FOTS coating but the settlement events on PEG and AWG were evaluated in the same way.

It is possible to interpret the recorded holograms without reconstruction because the movement of the spores can also be detected in the change of the interference fringes in the hologram. Settlement events can be detected in the hologram because the movement of a swimming spore suddenly stops. This is a fast possibility to obtain an overview of the recorded images. In Fig. 5.12 a hologram series is shown for the fluorinated (FOTS) surface. The series lasts 36:00 min in total (11,063 frames, or 13.5 Giga Byte (GB)). The experiment starts with the injection of the spores into the wet cell. This image is shown in the upper left corner, marked with the time label 00:00 min. The steel needle for the injection (black area at left side) and a clean surface are visible in this image. After the injection spores start to explore the surface. Settled spores are marked with colored points in the time series. To determine settled spores the analysis starts with the last image. The experiment is terminated by sucking out the spore suspension. As the surface was clean before the spores were injected, the four spores visible in the last image have settled during the experiment. The four spores adhered strongly enough to the surface to withstand the shear stress created by sucking out the suspension. These four positions are used to identify when a swimming spore has settled on the surface. This analysis is started at the last image and settled spores are tracked back in time until they start to move.

The spores marked red and green do not change their respective position on the surface from point in time 2:42 min after the spore injection. Both spores do not leave the surface until the experiment is terminated. Even though the spore marked green was on the surface for about 33 min it did not adhere strongly to the surface whereas the red marked spore withstand the shear stress and therefore is permanently adhered to the surface. The purple (since 4:07 min), blue (since 5:31 min), and orange (since 26:23 min) marked spores are, like the red spore (since 2:24 min), adhered to the FOTS surface strong enough to withstand the water shear stress. Consequently, spores do not have to settle when they establish a contact with the surface. In the hologram series (shown in Fig. 5.12) surface events occur and are not marked because the spores swim away after spending some time on the surface. With this analysis permanent settlement events are detected, but to obtain a deeper understanding of the settlement behavior, the holograms need to be reconstructed in order to observe these events with a better resolution.

In Sect. 5.2.2 the number of expected settlements was calculated. In Table 5.8, this numbers are compared to the observed events. For the FOTS surface the settlement expectation is 7.4 spores in the FoV after 40 min according to the kinetic study (see Table 5.8). In the holographic experiment four spores settled in the FoV (marked in Fig. 5.12) and two more spores (not marked in Fig. 5.12) are visible on the edges of the FoV. The observed settlement and the expected settlement are in good agreement. The settlement results for AWG and PEG (see Table 5.8) are also of the same order. On AWG two settlement events are observed in the FoV, whereas on PEG no settlement event could be recorded which is also expected.

Fig. 5.12 Hologram series for the FOTS surface. For all images the time in minutes is printed in the *upper left corner*. The first image (*upper left corner*, time: 00:00 min) is the injection of the spores. The steel needle is visible at the *left side* of the image (black) and the surface is clean. With elapsing time spores (marked with different colors) settle in the field of view. After 6:32 min (10,000 recorded frames) the frames are acquired with a lower frame rate. The experiment is terminated about 30 min later by sucking out the spore suspension. For four spores the released glue was already hard enough to withstand the shear stress created by sucking out the suspension, but the green spore is washed away. In total, during 36 min of the experiment, four spores settled in the field of view

The settlement analysis shows that the spores are "healthy". Therefore it is justified to regard the exploration behavior during the holographic recording as comparable to the standard assays.

Table 5.8 Comparison of the number of expected settlement events and of the observed settlement events on the investigated surfaces

Time (min)	PEG		AWG		FOTS	
	Expected	Observed	Expected	Observed	Expected	Observed
1	0.0		0.0		0.2	
5	0.0		0.1		0.9	
10	0.0		0.3		1.9	
20	0.0		0.5		3.7	
40	0.0	0	1.0	2	7.4	4(+2)

5.2.4 General Exploration Patterns

Before the general exploration patterns are discussed, the complete analyzed motion data is shown in Fig. 5.13. In this figure, the motion data from the surface (0 μm) until ≈ 800 μm in solution is shown for three independent experiments. The experiments (e.g. "AWG-A-*") are named according to system introduced in this chapter. For each individual surface experiment all analyzed different points in time are combined in only one diagram. The exact experimental details for each experiment can be found in the corresponding section (e.g. see Sect. A.1, Exploration behavior on AWG for the details of "AWG-A-*").

Figure 5.13 shows that spores move in an erratic, random motion and also on the surface no swarm behavior can be detected at this low spore concentration. The visible increase of size in the field of view (FoV) with greater distance to the surface is due to the holographic technique [12]. The motion data shows that the spore concentration is higher in the first 200 μm distances to the surface than further away. The effect—that microorganisms accumulate in the vicinity to the surface—is described in literature for bacteria [13, 14] and is also observed for spores in this study. This accumulation of spores shows that the surface has a great influence on the motility of spores. The motility in the bulk has already been discussed in Sect. 5.1. In the following section the motility within the first 200 μm above the surface is analyzed in detail.

In a previous work it has been shown that spores of the brown alga swim in different motion patterns [6]. One example is a light microscopy study on spores of the brown alga *Hincksia irregularis*. For *H. irregularis* five different swimming patterns are found. In Fig. 5.14 panel (a), a schematic sketch is shown for these swimming patterns ((A) *straight path*, (B) *search circles*, (C) *orientation*, (D) *gyration* and (E) *wobbling*). These motion pattern can be assigned to certain spore behavior (e.g. settlement, dying, exploration, swimming, etc.). These patterns will be referred to in the following as the "Iken pattern". The author claims that the motion analysis could facilitate a new antifouling bioassay by only determining the change in the ratio of RCD (rate of change direction) and SPD (swimming speed). It is shown that the RCD/SPD ratio changes when certain chemicals are released into seawater.

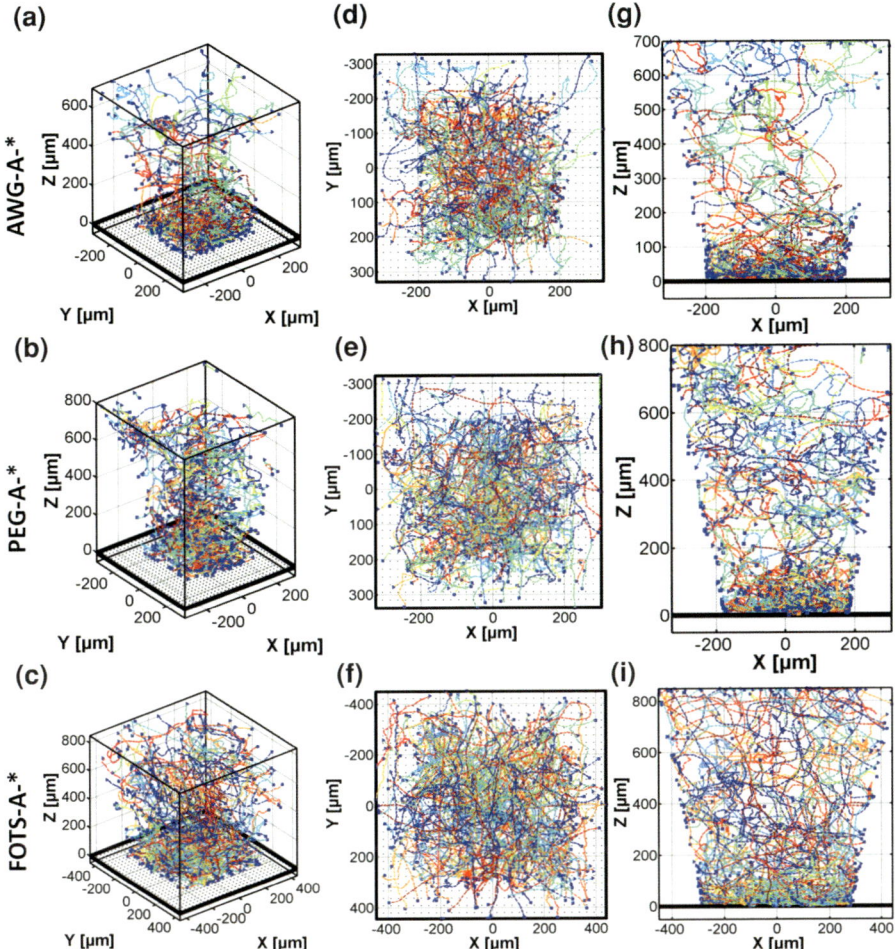

Fig. 5.13 3D rendered plots for spores collected at collection trip A. All analyzed trajectories from the surface (0 µm) until far into the solution (>700 µm) are shown: **a–c** 3D view; **d–f** xy view; **g–i** xz view

In Fig. 5.14 panel (b), the data of *Ulva* spores exploring a glass surface is shown [15]. The data was recorded in September 2006 and analyzed with the first trace determination software. It was possible to demonstrate the applicability of digital in-line holography to answer questions relevant to the understanding of *Ulva* spores surface exploration. The swimming patterns observed by Iken et al. [6] can also be found in the motion of *Ulva* spores and are described in greater detail in the following section. In the reconstructed images the occurrence of the "Iken patterns" can be linked to a certain distance to the surface which is not possible for the original 2D data. The motion data is classified in several motion patterns to describe the motility of *Ulva* spores in detail.

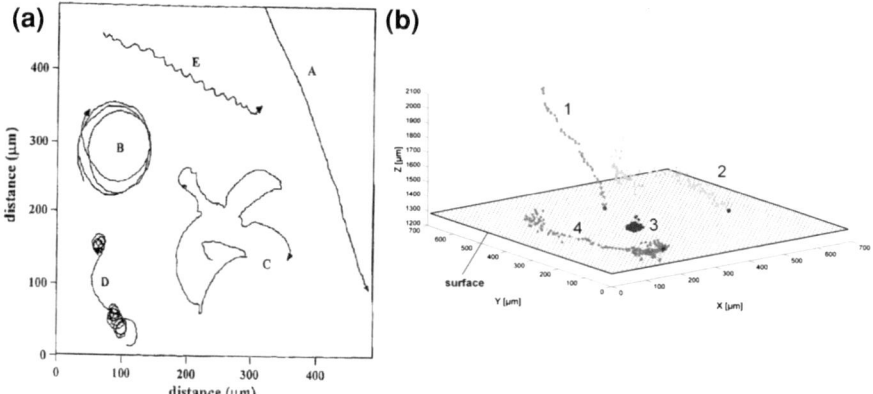

Fig. 5.14 a Swimming pattern of *H. irregularis* tracked by computer-assisted motion analysis: (A) *straight path*, (B) *search circle*, (C) *orientation*, (D) *gyration* and (E) *wobbling* [6]; **b** 3D exploration pattern for *Ulva* zoospores: (1) *straight path*, (2) *orientation*, (3) *search circle*, (4) *gyration* [15]

In analogy to the "Iken pattern" the following patterns are defined to describe the general motility of *Ulva* spores:

- *orientation* (Sect. 5.2.4.1)
- *wobbling* (Sect. 5.2.4.2)
- *gyration* (Sect. 5.2.4.3)
- *spinning* (former search circle [15]) (Sect. 5.2.4.5)
- *settlement* (Sect. 5.2.4.6)

Figure 5.15 shows a schematic overview of the identified motion patterns.

For a more detailed quantification of the surface exploration behavior two additional motion patterns are defined which combine the general motion pattern in a defined temporal frequency. These patterns are:

- *hit and run* (Sect. 5.2.4.4)
- *hit and stick* (Sect. 5.2.4.7)

It is important to note that the described patterns are not mutually exclusive for a recorded trajectory. Spores frequently swim in a combined pattern.

5.2.4.1 Swimming Pattern: Orientation

This pattern is defined for motion in solution. Figure 5.16 shows two (extreme) examples for the pattern. The swimming speed and the distance to the surface are the most important parameters to fit a spore trajectory to the swimming pattern. The pattern is only assigned if the spores swim faster (v_p of the complete trace) than 100 µm/s. The spore swims straight for a certain distance before it performs a turn and afterwards swims straight again. The distance that a spore swims straight

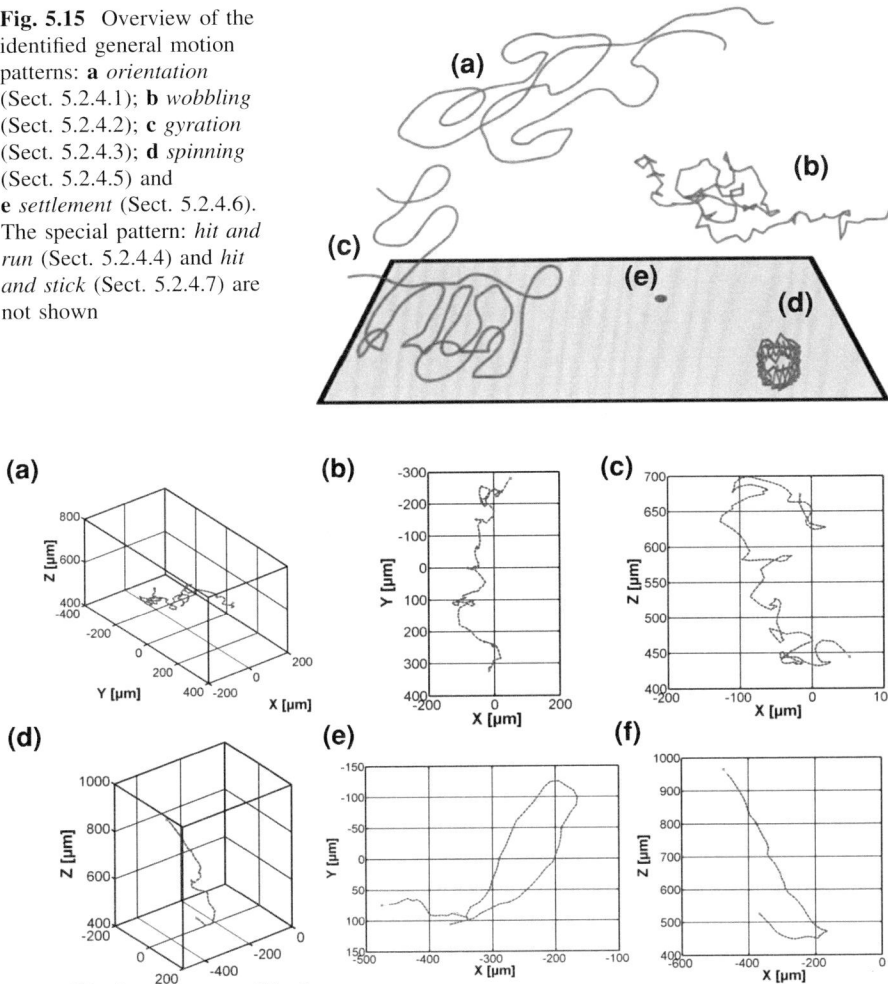

Fig. 5.15 Overview of the identified general motion patterns: **a** *orientation* (Sect. 5.2.4.1); **b** *wobbling* (Sect. 5.2.4.2); **c** *gyration* (Sect. 5.2.4.3); **d** *spinning* (Sect. 5.2.4.5) and **e** *settlement* (Sect. 5.2.4.6). The special pattern: *hit and run* (Sect. 5.2.4.4) and *hit and stick* (Sect. 5.2.4.7) are not shown

Fig. 5.16 Two examples for the swimming pattern orientation: **a, d** 3D view; **b, e** xy view; **c, f** xz view

before the next turn occurs is variable as visible in Fig. 5.16. All spores which are described in Sect. 5.1 as the fast spore fraction belong to this pattern. Therefore in Sect. 5.1.2.1 a detailed description of this swimming pattern is already provided.

5.2.4.2 Swimming Pattern: Wobbling

In Fig. 5.17, an example for the *wobbling* motion is shown. The spores are classified according to their swimming speeds. A spore is assigned to this pattern if v_p is significantly smaller than 100 µm/s. The spores assigned to the pattern often

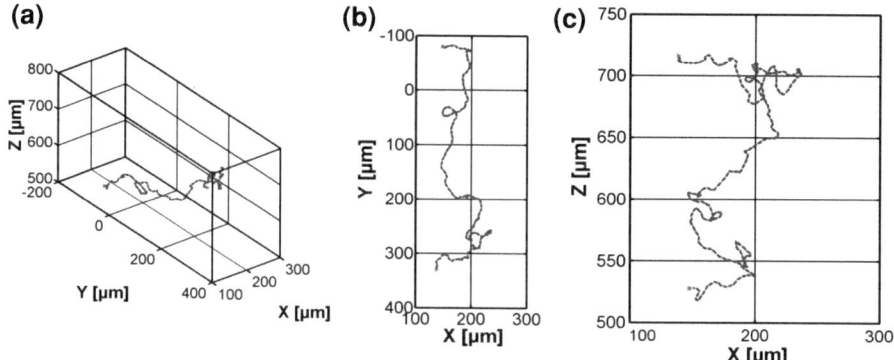

Fig. 5.17 Swimming pattern: Wobbling. This trace is discussed in Sect. 5.1.2.2 in detail: **a** 3D view, **b** xy view, **c** xz view

change their direction of movement and only swim in straight lines for very short distances. The spores described in Sect. 5.1.2.2 as the slow spore fraction are all assigned to this swimming pattern. Therefore, in Sect. 5.1.2.2 a detailed description of this swimming pattern is already provided.

In the study of *H. irregularis* by Iken et al. [6]. the appearance of this pattern is most probably explained with damaged spores. In Sect. 5.1.4, a detailed discussion of the origin (or kind, if gametes) of these spores is provided. The spores showing this behavior occur within the complete observation volume. Only two of the 147 wobbling spores within a distance of 200 μm from the surface swim with a strong preference towards the surface. Therefore these spores do not explore the surface actively.

5.2.4.3 Swimming Pattern: Gyration

For the exploration of surfaces this pattern is extremely important. The pattern occurs on all investigated surfaces but for each surface small differences within the pattern are observed. Therefore the pattern is discussed in detail separately for each investigated surface. This detailed analysis is provided in Sects. A.1.1, A.2.2 and A.3.1.

Swimmers are assigned to this pattern if surface contacts are observable and the trajectory does not belong to the *spinning* pattern. The *gyration* pattern can therefore be described as an *orientation* pattern with occurrence of surface contacts.

It is typical for the *gyration* pattern that the spore is not always in close contact to the surface. It swims towards the surface, sometimes it stays close to the surface for a certain time span, and then it moves away or starts to spin. Often the spore swims in wavelike motion (for the z-position, see Fig. 5.18, panel (c) black curve) relative to the surface. This behavior is a general searching behavior which is found for many other species in nature as well. The organism examines an area in

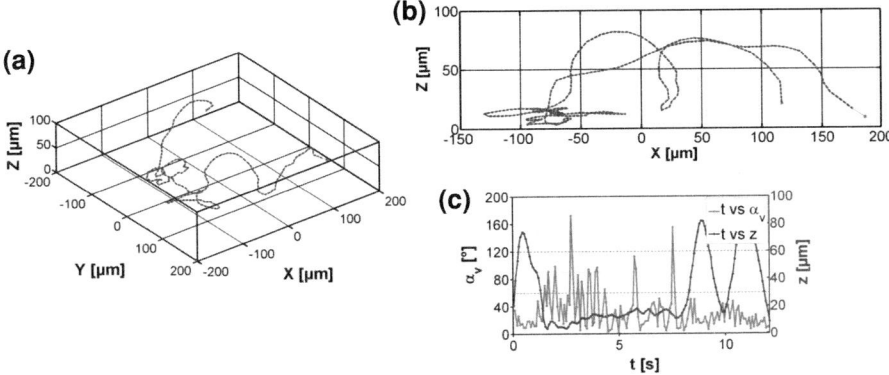

Fig. 5.18 Swimming pattern: *hit and run*. **a** 3D view, **b** xz view, **c** a_v versus elapsing time (*gray, left side*) and z-position versus elapsing time (*black, right side*)

detail, "decides" not to "like" it and then moves away from the surface to obtain an overview over the region and to start the next detailed exploration at a different position. A very common example for this search pattern are birds which search for nesting [16].

5.2.4.4 Swimming Pattern: *Hit and Run*

A special case of the *gyration* pattern is the *hit and run* pattern. The *hit and run* pattern is defined because it helps to describe the general exploration behavior and is useful for the comparison of the investigated surfaces. It is the extreme case of the *gyration* pattern and is defined as follows: A spore swims towards the surface and detaches from the surface after only a short surface contact. Afterwards no further surface contact is observed in the FoV for this spore (Fig. 5.19).

5.2.4.5 Swimming Pattern: Spinning

In an earlier light microscopy study [11] the authors described a motion in which the spore rotates in a rapid top-like configuration over the surface. During this motion the spore establishes contact between the apical "papilla" and the surface [11]. In high magnification and slow motion the rotation frequency was determined by taking the time of a specific feature of the spore (such as the eyespot- or a flagellar root) that needs to make a full rotation. The rotation speed for spores was determined to be 240 rpm. During the spinning the spore secretes a small amount of adhesive which is left behind when the spore stops spinning and swims away from the surface [11]. The *spinning* pattern of an individual organism can last for many minutes (observed for more than 5 min) without a significant change in the rotation frequency. Even after spinning for several minutes, the spore is still

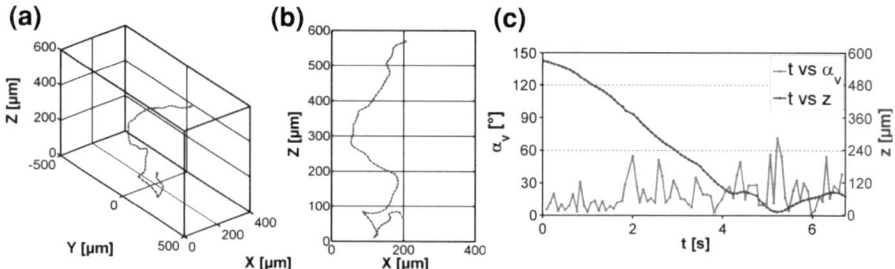

Fig. 5.19 Swimming pattern: *hit and run*, a 3D view, b xz view, c a_v versus elapsing time (*gray, left side*) and z-position versus elapsing time (*black, right side*)

able to swim away from the surface. Even if spinning does not necessarily lead to settlement, settlement only occurs out of the spinning motion. Typically before a spore settles the rotation frequency of the spinning motion gets slower until the spore shows lateral or "twitching" movements and finally permanently settles, not move anymore.

Figure 5.20 shows an example for this motion observed by the holographic tracking. The center of mass of the spore is shown as a black dot. To obtain an image for the dimensions a sketch of a spore is also included in the figure. The motion data was acquired with a frame rate of 11 Hz and therefore the spore position is only captured at three positions within a circle until the spore has performed a complete turn. Due to the low acquiring frame rate in the holography motion study it is not possible to proof the observation of the light microscopy study [11] that the spore rotates on the surface. It is also possible to interpret the holographic data in the way that the spore is pinned at the apical "papilla" and "twitches" around this position on the surface. However, since this motion is a 2D event it is studied in greater detail (higher resolution, higher frame rate) by standard light microscopy so that the motion is named and described in analogy to the previous study [11]. The motion is characterized in the following by the angular frequency (af) $\left(\frac{\beta}{dt}\right)$ and the radius (ra) of the circle. In Fig. 4.18 a definition of the radius (ra) and β is provided. To compare the motion with the motion observed in light microscopy the rotation frequency (rf) is also calculated.

The *spinning* pattern is not observed on the PEG surfaces but on AWG and FOTS coating. On AWG and FOTS the pattern is similar. However, it is discussed later in detail for each surface separately.

5.2.4.6 Settlement

Settlement occurs out of the *spinning* pattern. While spinning the spore releases its adhesive. This process is studied better and in more detail in the earlier light microscope study [11]. In Sect. 5.2.3 the details for the settlement events on the FOTS surface are described. Settlement is an important process in the spore lifecycle,

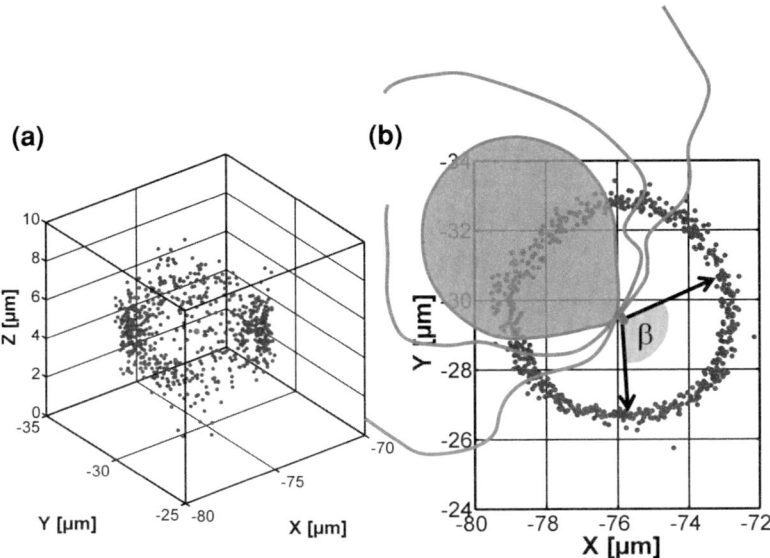

Fig. 5.20 Swimming pattern: *spinning*. **a** 3D view and **b** xy-view for the motion pattern. Each blue dot describes the center of mass of the spore. To obtain a feeling for the dimensions a sketch of a spore is included in the figure. The spinning motion is characterized in the following by the angle β and the radius of the circle

but the focus of this work is on the initial approach to the surface and exploration behavior which leads to settlement. From a motility analysis point of view the settlement event cannot quantitatively be studied as it is simply a static point.

5.2.4.7 Swimming Pattern: *Hit and Stick*

The *hit and stick* pattern is unique for FOTS and is therefore not defined as a general motion pattern. In Sect. A.3.1.1 the pattern is described in detail. It can be divided into four parts: (i) approach, (ii) *sticking*, (iii) *spinning*, (iv$_a$) detachment or (iv$_b$) settlement. The pattern combines general motion patterns (*orientation*, *spinning* and *settlement*) but it is defined as an individual pattern because many spores swim exactly according to the above defined temporal order of motion patterns. On none of the other studied surfaces (AWG, PEG) a swimming motion which is similar to *sticking* phase in the *hit and stick* pattern is observed. Figure 5.21 shows an example for the sticking phase.

5.3 Summary of the Results of the Surface Exploration

In the following the exploration behavior in the vicinity to the studied surfaces is demonstrated in a short and concise manner. In the appendix (Sects. A.1–A.3) a complete analysis is provided to address arising questions in detail. The analysis

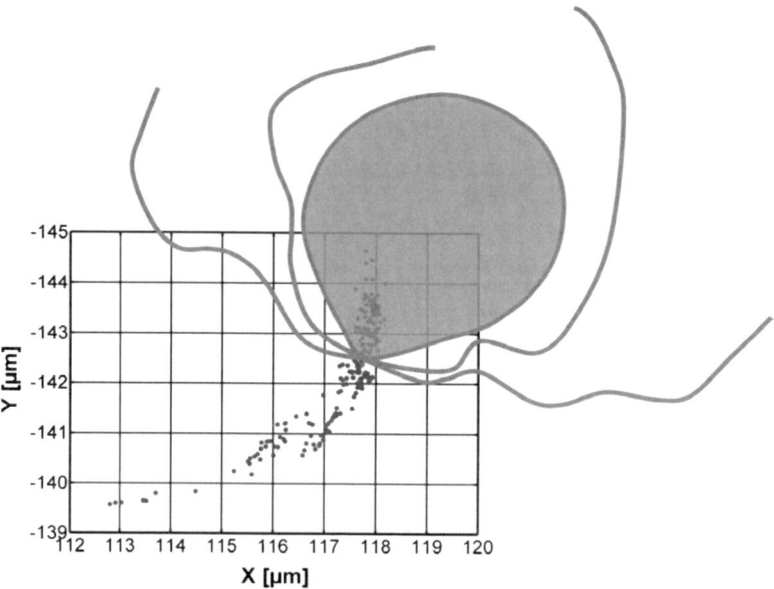

Fig. 5.21 Example for the *sticking* phase during the *hit and stick* pattern. A sketch of the spore is included to obtain a feeling for the dimension of the motion during the sticking phase. The *black dots* denote the center of mass of the spore

for each surface follows the same structure so that each section can be understood independently. For every surface, first, the motility in vicinity to the surface is discussed by showing exemplary trajectories and classification using the swimming patterns described in Sect. 5.2.4. Subsequently the general motility is discussed by the velocity histograms and by the detailed analysis of the motility parameters (velocity, α_v and α_z). Section A.1 discusses the observed behavior in the vicinity to AWG, Sect. A.2 to PEG and Sect. A.3 to FOTS. The comparison of the behavior in respect to the surfaces mentioned above is discussed in Chap. 6.

5.3.1 General Observations on Motility for All Investigated Surfaces

- It is observed that the spores belonging to the *wobbling* pattern do not actively explore the surface whereas the spores assigned to the *spinning, gyration* and *orientation, hit and run*, and *hit and stick* patterns search for a place to settle.
- To study the behavior of spores searching for a surface position suitable for settling only a subclass of the defined motion patterns is necessary. These are *gyration, hit and run, orientation*, and the approach and detachment assigned to the *hit and stick* pattern. They provide the major information and are therefore referred to in the following as the "active searching motion" (ASM). In contrast,

spores performing the *spinning* pattern locally probe the surface at a fixed surface position and do therefore not provide information on the surface approach.

5.3.2 Summary of Results for the Exploration Behavior on AWG

- In the FoV two settlement events are observed after 40 min. Even if the number appears to be small it comes up to the expectations for the total amount of settlement during the holographic recording. The occurrence of settlement allows the conclusion that the spore behavior is comparable to the observed activity in the standard assays. For details see Sect. 5.2.3.
- The spores accumulate in vicinity (0–200 µm) to the surface (see Fig. A.15).
- The spores swim slower if they are close to the surface. At a distance of 120 µm from the surface the swimming speed starts to decrease (see Fig. A.17). Furthermore, in a distance of 60–200 µm from the surface the spores swim faster towards than away from the surface (see Fig. A.18).
- The swimming speed drops significantly if the spores swim at distance of 0–30 µm from the surface. This drop in velocity can be correlated to the interaction strength of spores with the surface.
- The $\bar{\alpha}_v$ distribution changes significantly if a spore swims close enough to the surface. This means that a spore changes its swimming direction more frequently than in the bulk. This increase in $\bar{\alpha}_v$ is observed at a distance of 60 µm from the surface (see Fig. A.16).
- The $\bar{\alpha}_z$ distribution shows that the spores have a preference to swim towards the surface in the bulk (240–720 µm). In vicinity to the surface (0–240 µm) no preference in the swimming direction is observed (see Fig. A.16).
- The trend for the surface exploration is reproduced in two individual experiments (AWG-I-A-* and AWG-II-B-*) for spores harvested at different days.

5.3.3 Summary of Results for the Exploration Behavior on PEG

- No settlement is observed on the coating which illustrates the inhospitability of the PEG coating (see Sect. 5.2.3).
- The center of mass of the spore body—except for the spores analyzed within the trajectory "PEG-Un-1"—are not observed in the surface plane. The closest distance of the center of mass of the spore body is found 5 µm from the surfaces. However, for most spores a parallel motion along the surface or movement away

from the surface occurs if the center of mass of the spore body is detected 15 μm from the surface.

- Spores accumulate in the vicinity (0–200 μm) to the surface.
- The velocity of the swimming spores in vicinity to the surface (up to 180 μm) is slightly smaller than in the bulk. There is no significant difference between the spores which swim towards or away from the surface. This means that the spores have a small interaction strength with the surface.
- The $\bar{\alpha}_v$ distribution strongly increases starting at a distance of 180 μm from the surface. A considerable difference for the $\bar{\alpha}_v$ distribution is observed for the spores approaching and swimming away from the surface.
- Up to a distance of 90 μm from the surface a small preference is observed for the spores to swim towards the surface ($\bar{\alpha}_z$ distribution $>90°$).
- The trace "PEG-Un-1" is the only recorded spore which explores the surface. Its swimming characteristic is somewhere between the *gyration* and *spinning* pattern (see Sect. A.2.1.3).

5.3.4 Summary of Results for the Exploration Behavior on FOTS

- Six settlement events are observed in the FoV (see Sect. 5.2.3).
- The *hit and stick* pattern, which is unique for FOTS, is observed.
- The exploration behavior of the surface changes with elapsing recording time. In the beginning (FOTS-A-1) the surface is explored via the *hit and stick* pattern, while the *gyration* pattern is not observed. In FOTS-A-3 only the *gyration* pattern is observed whereas the *hit and stick* pattern does not occur anymore. In FOTS-A-2 both patterns are found and therefore this dataset set the point in time of the changes in the exploration behavior (see Sect. A.3.2).
- The spore accumulation in the vicinity to the surface changes for the individual experiments. In the beginning (FOTS-A-1) no accumulation is observed. With elapsing time the typical accumulation in the vicinity (0–220 μm) to the surface is observed (FOTS-A-3) (see Fig. A.50).
- The $\bar{a}_z$ distribution also changes during the experiment. In the beginning (FOTS-A-1) a strong preference is observed to swim towards the surface. This flow of spores towards the surface is observed for the complete observation volume ($>1,000$ μm) and is significantly stronger pronounced than for AWG and PEG. Later (FOTS-A-3) this preference is lost nearly for the complete observation volume and is similar to the other surfaces (see Figs. A.52 and A.54).
- In FOTS-A-1 the spores approach the surface steeper than in the following experiment. For the detachment the opposite trend is observed. In FOTS-A-1 the spores leave the surface with a small angle to the surface normal. In the later experiments the detachment angle increases. In FOTS-A-1 the approach angle

and the detachment angle are significantly different. This difference is lost during the experiment (see Figs. A.52 and A.54).

- The approach angle for all observed trajectories assigned to the *hit and stick* pattern is 50° while the detachment angle is 21°. The spores approaching the surface are faster (223 ± 48 μm/s) than the spores leaving (168 ± 38 μm/s) the surface (see Table A.8).
- The velocity distribution changes during the experiment. For FOTS-A-1 the spores leaving the surface are faster than those approaching the surface. A dip in the velocity distribution of the spores approaching the surface is observed at a distance of 90 μm from the surface. This peak coincides with the peak in the $\bar{a}_z$ distribution. This means that the spores perform a turn at a distance 90 μm from the surface and swim towards the surface (see Figs. A.53 and A.55).
- In experiment FOTS-A-2, -3 the spores leaving the surface are significantly slower than the spores approaching the surface. At a distance of 240 μm from the surface the swimming speed starts to get slower for both spore fractions (towards and away from the surface) (see Fig. A.55).

References

1. Wikipedia, Web Page, Maxwell–Boltzmann distribution, http://en.wikipedia.org/wiki/Maxwell-Boltzmann_distribution. Accessed: 27 Oct 2009
2. P. Hegemann, Planta **203**, 265–274 (1997)
3. D.R. Mitchell, J. Phycol. **36**(2), 261–273 (2000)
4. K. Wakabayashi, S.M. King, J. Cell Biol. **173**(5), 743–754 (2006)
5. I. Inouye, T. Hori, Protoplasma **164**(1–3), 54–69 (1991)
6. K. Iken, C.D. Amsler, S.R. Greer, J.B. McClintock, Phycologia **40**(4), 359–366 (2001)
7. A. Rosenhahn, T. Ederth, M.E. Pettitt, Biointerphases **3**(1), IR1–IR5 (2008)
8. S. Schilp, A. Rosenhahn, M.E. Pettitt, J. Bowen, M.E. Callow, J.A. Callow, M. Grunze, Langmuir **25**(17), 10077–10082 (2009)
9. J.A. Finlay, S. Krishnan, M.E. Callow, J.A. Callow, R. Dong, N. Asgill, K. Wong, E.J. Kramer, C.K. Ober, C.K. Ober, Langmuir **24**(2), 503–510 (2008)
10. I. Thome, Einfluss der Oberflächenchemie und -Morphologie auf die Besiedelung von Oberflächen durch maritime Organismen, Diploma Thesis, Ruprecht-Karls-University of Heidelberg, Heidelberg, 2009
11. M.E. Callow, J.A. Callow, J.D. Pickett-Heaps, R. Wetherbee, J. Phycol. **33**(6), 938–947 (1997)
12. J. Garcia-Sucerquia, W. Xu, S.K. Jericho, M.H. Jericho, P. Klages, H.J. Kreuzer, Proc. SPIE Int. Soc. Opt. Eng. **6027**, 60272H/1–60272H/8 (2006)
13. L. Rothschild, Nature **198**(4886), 1221–1222 (1963)
14. A.P. Berke, L. Turner, H.C. Berg, E. Lauga, Phys. Rev. Lett. **101**(3), 038102(1)–038102(4) (2008)
15. M. Heydt, A. Rosenhahn, M. Grunze, M. Pettitt, M.E. Callow, J.A. Callow, J. Adhesion **83**(5), 417–430 (2007)
16. P.J. Li, T.E. Martin, AUK **108**(2), 405–418 (1991)

Chapter 6
Discussion of the Motility of *Ulva* Zoospores in Vicinity to Surfaces

In this chapter the surface interactions of spores are compared. Therefore the motility of *Ulva* spores in solution (Sect. 5.1.4) is set into context with the motility observed in vicinity of the surface. Although multiple measuring trips were conducted throughout the course of this thesis, only results from the last measuring session are shown. The reason being, that for the previous trips the experiment was in an optimization process to obtain a convection free environment, greater magnification, and better image quality. However, the major results (*spinning* and settlement on AWG [1], no settlement in the FoV on PEG, and the *hit and stick* pattern on FOTS) are observed and reproduced for all the experiments.

6.1 Occurrence and Time Evolution of the Exploration Behavior in Vicinity to Different Surfaces

For the first time the spore motility in vicinity of various surfaces is determined and analyzed in 3D. The observed motility is different on the investigated surfaces and can be correlated to the attractiveness of a surface. The latter describe how fast the surface is colonized by the organisms. This settlement kinetic is determined by spore settlement assays which are described in Sect. 4.6.

To characterize the spore behavior different motion patterns are defined and described in detail in Sect. 5.2.4. In Appendix, Sects. A.1.1, A.2.1 and A.3.1 the observed movement characteristics within these patterns are described for each surface. The defined motion patterns encode different surface interaction. The shortest surface interaction time is observed for the *hit and run* pattern whereas the *gyration* pattern describes a longer surface interaction time. While swimming in a *gyration* fashion the spore spends a reasonable amount of time on the surface but moves around steadily. The next longer spore surface interaction time is observed within the *spinning* pattern and the longest spore–surface interaction time are found during the *sticking* phase of the *hit and stick* pattern.

M. Heydt, *How Do Spores Select Where to Settle?*, Springer Theses,
DOI: 10.1007/978-3-642-17217-5_6, © Springer-Verlag Berlin Heidelberg 2011

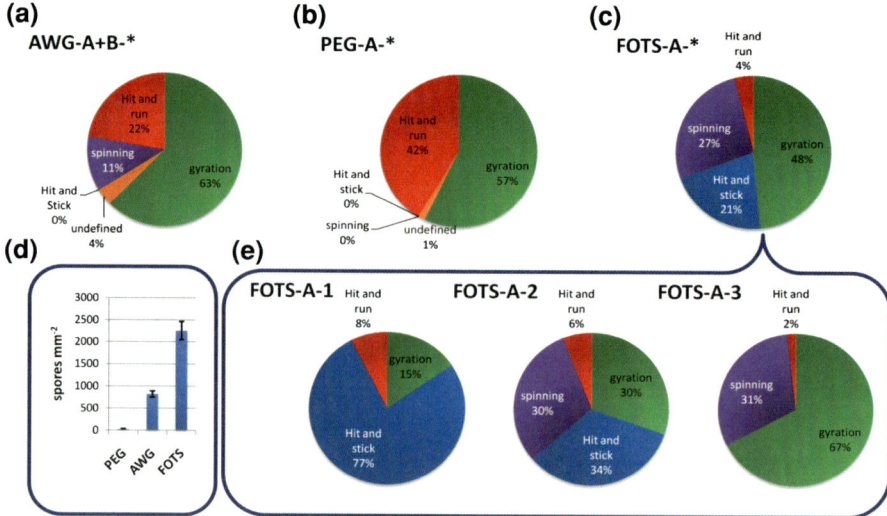

Fig. 6.1 Summary of the observed motion patterns for the investigated surfaces (**a–c**). **d** Spore settlement amount after 45 min observed in a standard AMBIO settlement assay [2]. **e** For FOTS the exploration behavior changes with elapsing time so that the individual experiments are shown in detail

Figure 6.1, panels (a–c) gives a summary of the occurrence of the observed motion patterns on each surface. The amount of settlement on the used surfaces is studied by a standard AMBIO settlement assay [2]. The result of this assay is shown in Fig. 6.1, panel (d). After 45 min of spore incubation almost no spores have settled on PEG (27 ± 11 spores mm^{-2}), whereas quite a number of spores select AWG as suitable for settlement (820 ± 67 spores mm^{-2}). The highest amount of settlement is observed on FOTS (2255 ± 207 spores mm^{-2}).

For the intermediately attractive AWG surface the *gyration* pattern is the most dominant motion pattern (63%, see Fig. 6.1, panel (a)), whereas 11% of the spores belong to the *spinning* pattern. For these spores the surface appears to be worth for a more detailed exploration. On average every fifth spore (22%) finds the surface as not suitable for settlement after the first contact (*hit and run* pattern).

In comparison to AWG the situation on the unattractive PEG coating (panel (b)) is different. The amount of spores identifying the surface as not suitable for settlement after the first contact is twice as high as on AWG (42% *hit and run*). Nevertheless, the amount of spores assigned to the *gyration* pattern is still of the same order (57%) compared to AWG (63%). However, the most striking difference between the surfaces is that no *spinning* event is observed on PEG.

It is observed that the settlement on FOTS is fairly high in respect to the other surfaces. This high amount of fouling can also be seen in the distribution of the motion patterns. The percentage of the *hit and run* pattern is small (4%, panel (c)) in comparison to AWG (22%) or PEG (42%) while the percentage of the *spinning* pattern is significantly higher on FOTS (27%) than on AWG (11%) or PEG (0%). Furthermore, on FOTS a spinning phase is involved in each *hit and stick* pattern (21%) and therefore an effective percentage of 48% spinning is observed for all analyzed spores on FOTS. The temporal change in the occurrence of the exploration pattern on FOTS (panel (e)) is discussed in Sect. 6.3 in which the exploration pattern on FOTS, especially the *hit and stick pattern,* is discussed in detail.

The occurrence of the *spinning* pattern (FOTS: 48% (21% + 27%), AWG: 11%, PEG 0%) goes along with the determined settlement by the AMBIO standard settlement assay. But also the amount of the *hit and run* pattern (FOTS: 4%, AWG: 22%, PEG: 42%) correlates with the assay result. Therefore—shown for the first time—the spore exploration behavior in vicinity of a surface can be correlated to the fouling rate of irreversible adhesion during the first 40 min of contact time determined by the AMBIO standard settlement assay. The occurrence distribution of the motion patterns is distinctive for each investigated surface and encodes the expectations for settlement on the surface.

Furthermore, the analysis of a few short sequences with a time span of 30 s out of the recording within 10 min of spore incubation are sufficient for the 3D motion analysis to answer the question whether a surface is attractive (early occurrence and high percentage of *spinning*, e.g. FOTS) or unattractive (no spinning, high percentage of *hit and run*, e.g. PEG). Especially, the first 3 min are sensitive and thus predict the outcome of the standard AMBIO settlement assay. This result can envisage for future high throughput screening technique based on the motion analysis to anticipate the antifouling performance of a coating as measured in the assays.

6.2 Deterrent Properties of the PEG Surface

In contrast to the standard AMBIO settlement assay which only observes the total number of settled spores, the 3D motion pattern study analyzes the settlement behavior itself. This knowledge can provide information to understand the mechanism causing a surface being attractive or unattractive for spore settlement.

It is necessary for a spore to spin in order to form an adhesive pad before it is able to settle [3–5]. During the spinning the spore establishes contacts between the spore body and the surface. No spinning event is observed on PEG within the observation time and in the FoV. This observation corresponds with the low settlement found in the standard AMBIO settlement assay, but why does the pattern not occur? To answer this question the motion pattern on PEG is discussed in detail.

In the discussion, only spores assigned to ASM (Active Searching Movement, for a definition see Sect. 5.3) are considered. When the trajectories within the vicinity of the surface are compared for AWG and PEG (Fig. 6.2, panels (a, b)) the center of mass of the spore body is not detected within the surface plane on PEG, whereas on AWG contacts between the cell bodies and the surface are observed.

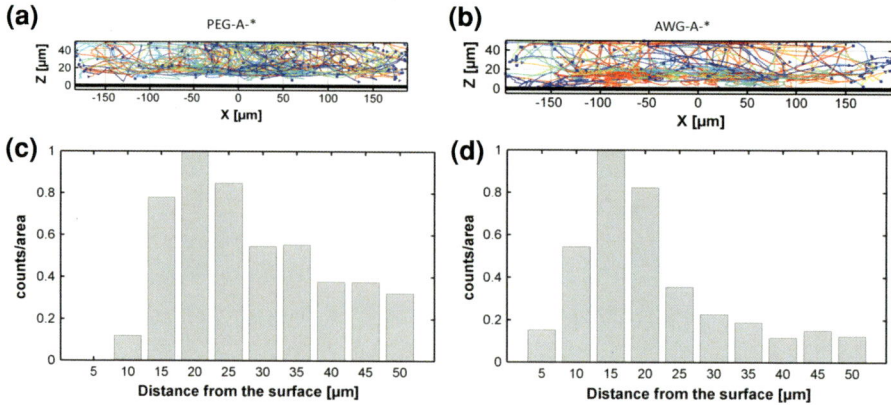

Fig. 6.2 Spore distribution close to PEG (**a, c**) and AWG for the spores assigned to ASM (Active Searching Movement, see Sect. 5.3). **a, b** Observed trajectories and **c, d** relative spore distribution

The spore distribution can be studied in more detail in the histogram shown in Fig. 6.2, panels (c, d). For both surfaces the highest spore concentration is observed at a distance of 15–20 μm from the surface, but aside from the maximum the shape of the spore distribution is different for the investigated surfaces. The distribution for AWG is nearly symmetrical around the maximum. The spores assigned to ASM prefer to stay at this distance and swim along the surface. The probability for a spore observed at a distance of 15 μm to swim towards or away from the surface is similar for both cases. In contrast, for PEG the spore distribution is asymmetrical around the maximum. No spores assigned to ASM are observed in the section 0–5 μm (depletion volume) and in the section 5–10 μm significantly less spores are observed compared to AWG. In contrast to AWG the spore concentration on PEG is higher for greater distances (>20) from the surface. This means that the spores rather swim away than towards to the surface if they are at a distance of 15–20 μm from the surface. In the following the different spore motilities near the surfaces are characterized by the swimming parameters: $\bar{a}_v$, $\bar{a}_z$ and v_m shown in Fig. 6.3.

In Sect. 4.4 the angle $\bar{a}_v$, which describes the direction of the movement (e.g. straight or circular), and $\bar{a}_z$, which describes the movement towards or away from the surface, are explained in detail. The mean swimming velocity is abbreviated by v_m. Figure 6.3 shows the motion parameters $\bar{a}_v$, $\bar{a}_z$ and v_m distribution for spores assigned to ASM and for PEG and AWG. The distribution is shown for a span from the surface up to 250 μm into the water column. In all panels the blue marked line describes all analyzed spores, whereas the red marked line represents the sub spore fraction swimming towards the surface and the green marked line shows the sub spore fraction swimming away from the surface.

Figure 6.3, panels (a, b) shows $\bar{a}_v$ distribution for PEG and AWG. The general trend for both distributions is similar. In the water column at a distance bigger than

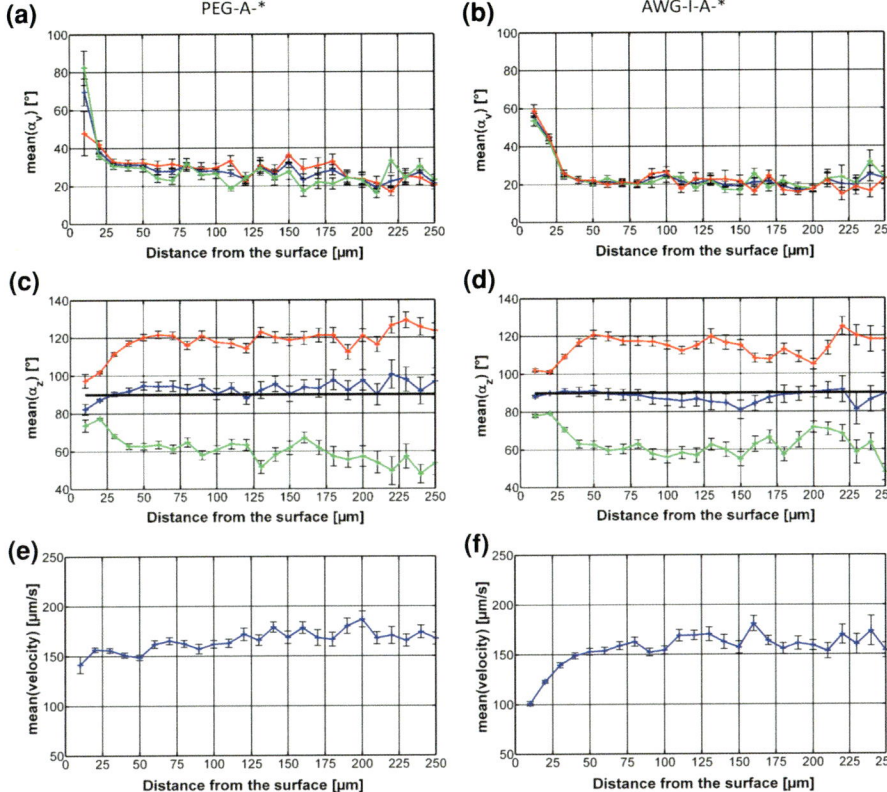

Fig. 6.3 Comparison of the exploration behavior on PEG and AWG. Only the spores assigned to ASM are shown. **a, c, e** the behavior for PEG, whereas in **b, d, f** the behavior for AWG are illustrated. **a, b** $\bar{a}_v$ distribution, **c, d** $\bar{a}_z$ distribution, **e, f** v_m distribution and panels. In **a–f** the line color encodes the following: *red line*, towards the surface; *blue line*, all spores; and *green line*, away from the surface

50 μm to the surface the value of $\bar{a}_v$ scatters around a certain value. The value is for PEG 27 ± 2° and for AWG 20 ± 2°. At a distance of 50 μm from the surface the distribution changes and increases significantly. To study the motility causing the depletion volume the distributions in the first two bins (0–10; 10–20 μm) are important. For AWG no difference between the spore fractions swimming towards (red) and away (green) from the surface is observed. At the section 0–10 μm the value of $\bar{a}_v$ is 58 ± 5°. In contrast on PEG the shape of the $\bar{a}_v$ distribution is different. There is a significant difference between the spore fractions swimming towards ($\bar{a}_v = 48 \pm 12°$, red curve) and away ($\bar{a}_v = 83 \pm 10°$, green curve) from the surface. The value for the spores swimming towards the surface is within the error in the same range as observed on AWG, but the value of $\bar{a}_v$ observed for the fraction swimming away from the surface is significant higher. Before the meaning of this observation for $\bar{a}_v$ is discussed the $\bar{a}_z$ distributions for PEG and AWG are explained (Fig. 6.3, panels (c, d)).

As for $\bar{a}_v$ in general the distribution for both surfaces is similar. For all analyzed spores (blue curve) the value of the $\bar{a}_z$ distribution scatters around 90° for the complete volume. With greater distance to the surface the statics of the observed spores are smaller and therefore the error bars get bigger. Therefore, no trend is observed of more spores swimming away or towards the surface. The $\bar{a}_z$ distribution is isotropic if the mean angle is 90°, and for the half space which describes the motion away from the surface (that corresponds to values in α_z between 0° and 90°) the value for an isotropic distribution is 56°, for the fraction swimming towards the surface it is 124° (see Sect. 4.4, Fig. 4.17).

Depending on the distributions in the half spaces the movement on AWG and PEG is isotropic for a distance greater than 50 µm ($\approx 60°$ for away and $\approx 120°$ for towards). There is no significant difference in the shape of the distribution for the movement away or towards the surface. Closer than 50 µm to the surface the value of $\bar{a}_z$ approaches 80° for the spores swimming away and 100° for the spores swimming towards the surface. This change in the value of $\bar{a}_z$ towards 80° (or 100° respectively) means that the spores do not move isotropic but rather along the surface.

Even if the spores swim on both surfaces parallel to the surface, close to the surface the spores on AWG are slower than the spores observed on PEG. In the section 0–10 µm from the surface the average swimming speed on AWG is decreased by 38% whereas on PEG the velocity only decreases by 17%. This difference in the decrease means that the surface–spore interaction is much weaker on PEG than on AWG. The difference for the velocity distributions is not only observed in the section 0–10 µm but also in the section 10–20 µm. On AWG the interaction is significantly stronger than on PEG resulting in the slowdown. At greater distances the difference in the slowdown between the surfaces is not obvious anymore (section 20–30 µm decrease: AWG 13%, PEG 12%).

A possible explanation for this observation is that the spore is able to interact with the surface by its flagella before the cell body gets in contact with the surface. The flagellar beating pattern of *Ulva linza* zoospores is not jet studied in detail. However, the swimming performance of other algae and of *Ulva* gametes (which only have two instead of four flagella) is studied by Inouye et al. [6]. The flagella arrangement during swimming is shown in Fig. 6.4 and explained in detail in Sect. 3.1. The complete length of the *Ulva linza* flagellum is 15 µm and the flagella are hold back in a cruciate pattern during forward swimming. During the fusion of gametes the first contact between the cells is established at the tips of the flagella [7]. This means that the spores can use the flagella not only for swimming but also to start the signaling process which triggers cell fusion.

For the spores assigned to ASM the highest spore density is observed at a distance of 15 µm from the surface (shown in Fig. 6.2). For most spores a swimming motion parallel to the surface is observed at this distance. Due to the cruciate arrangement of the flagella, at least one flagellum has to get very close to the surface during the beating pattern while swimming parallel to the surface. If the spore swims closer to the surface the flagella could come in contact with the surface during the beating pattern. This contact results in the observed drop in the

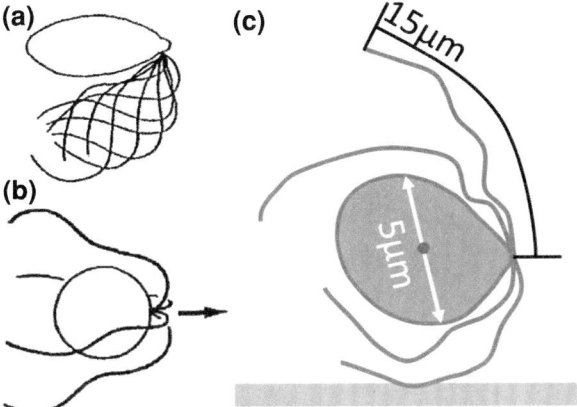

Fig. 6.4 Sketch to illustrate the flagella arrangement and the dimensions of a swimming *Ulva* spore. **a** Flagellar beating pattern of gametes from the alga *Ulva Bryopsis maxima* which has two flagella [6]; **b** lagella arrangement of *Prasinopapilla vacuolata* during normal forward swimming holding their flagella back in a cruciate profile [6]; **c** sketch of a *Ulva* zoospore with the marked dimensions for the cell body and the flagella

swimming speed observed for AWG (Fig. 6.3) and FOTS-A-3 (discussed in the Sect. 6.3, Fig. 6.7). On PEG this drop in the velocity is not as strong, which leads to the conclusion that the interaction between the flagella and the surface is weaker. Furthermore, only on PEG the center of mass of the spore body is not observed within the surface plane (see histogram Fig. 6.2). The observed $\bar{a}_v$ distribution on PEG (Fig. 6.3, towards: $\bar{a}_v = 48 \pm 12°$, away: $\bar{a}_v = 83 \pm 10°$) can be interpreted the way that the spores swim away from the surface immediately when getting closer than 10 μm to the surface. This movement leads to the observed depletion layer next to the surface. The exploration behavior has to be due to the properties of the PEG coating which the spore is probably able to examine once the flagellum get in contact with and the surface. PEG is a highly hydrated polymer and its anti-adhesive properties as well as the settlement inhibition are based on steric repulsion of the highly hydrated and loosely packed chains forming a diffuse interphase [8–11]. The tip of the flagellum is too large and complex in order for the initial concept of steric repulsion being exclusively able to explain the interactions between the flagellum and surface. Nevertheless, the first contact between the flagellum and the PEG interface also occurs on a molecular level. Therefore, it appears to be possible (especially, if it is envisioned that the flagellum moves in a wave like fashion) that, at the first contact, the flagellum does not stick to the interface but rather slides along the interface. This potential sliding of the flagellum results in a weak surface interaction which has a deterrent effect on the exploration behavior.

Thus, the fact that no *spinning* occurs on PEG is connected to the fact that the spores swim away from the surface if they are getting closer than a distance of 10 μm to the surface. Probably the spores sense the surface properties of PEG with the tips of their flagella and identify the surface as not suitable for settlement and thus swim away. To our knowledge this effect of active response to the properties of the PEG interface is shown for the first time for relatively large (5 μm) swimming microorganisms. Furthermore, the flagellum–surface interaction is significantly smaller on PEG than on AWG. This interaction is observed by the velocity slowdown on both surfaces. The interaction between flagella and the AWG surface must be higher than on PEG, because the slowdown in the swimming speed is already observed at a distance where contact of the spore body with the surface is observed. Based on this discussion it is possible to postulate the hypothesis, that a spore needs a sufficient strong flagella surface interaction to establish a spore body surface contact.

With elapsing time, even on PEG, a few spores are observed to settle on the surface (30 spores mm^{-2}). These spores are able to overcome the repulsion of the PEG coating. The trajectory PEG-Un-1 is the first spore after the injection into the wet cell observed to overcome the deterrent effect of the interface.

Schilp et al. showed that, even if a spore achieves to actively overcome the steric repulsion and settles on the surface, the released glue cannot penetrate the coating (because of its anti-adhesive properties (steric repulsion)) and it is only loosely attached [10]. Thus the spore can be rinsed easily from the surface [10]. Whether a spore "knows" in advance that its glue will not stick to the surface and therefore selects the surface as not suitable for settlement is not known and appears to be unlikely. It is more likely that the spore "interprets" the PEG coating with its flagella as a place not suitable to settle. Interestingly, on an EG_6 coating which is, as PEG, a protein resistant surface the spores are also not able to adhere to the surface. Nevertheless the observed amount of initial settlement on EG_6 is high in contrast to the very low settlement on PEG [12]. The protein resistance which explains that the spores cannot stick to the surface is caused by different surface properties on PEG and EG_6: EG_6 is protein resistant because of a water layer strongly bound to the surface, whereas PEG is protein resistant because of steric repulsion. This difference in surface properties leads to the fact that while both surfaces are anti-adhesive and protein-resistant, PEG additionally shows a deterrent effect on the exploration behavior of spores. Therefore PEG fulfills all necessary requirements for an antifouling coating, whereas EG_6 is "only" a very good fouling release coating.

6.3 The *Hit and Stick* Pattern and Its Importance for the Observed High Amount of Settlement on FOTS

The observation unique for the FOTS surface is the *hit and stick* pattern. Shortly after the spores' injection into the wet cell the first spores arriving at the surface are trapped at the interface. The *hit and stick* pattern is only assigned to spores for which a *sticking* phase is observed in the trajectory. The *hit and stick* pattern only occurs in the first 3 min after the injection. At a later point in time the exploration pattern in vicinity to FOTS is similar to the observed motility on AWG. Nevertheless, the number of settled spores observed after 45 min by the AMBIO settlement assay is high on the FOTS coating in comparison to AWG. Therefore the short trapping phase in the beginning has to have an influence on the fouling rate on FOTS because otherwise the amount of fouling on AWG and FOTS should be fairly similar.

The change in the exploration behavior is observed for many exploration parameters (α_v, α_z velocity, spore enrichment close to the surface) but is first discussed with the help of the relative occurrence of the motion patterns shown in Fig. 6.5. For FOTS-A-1 the most dominant pattern is the *hit and stick* pattern (77%). Experiment FOTS-A-1 starts 0:29 min after the injection and last until 1:24 min. For the first 30 s of FOTS-A-1 the *hit and stick* pattern is nearly the exclusively observed motion pattern next to the surface. With elapsing time the other surface exploration patterns also occur, while the *hit and stick* pattern gradually vanishes. On FOTS spinning occurs at any point in time, because each *hit and stick* pattern includes a spinning phase. Due to the *hit and stick* pattern the *spinning* pattern is populated to a higher degree than for the other surfaces. In comparison on AWG the first spinning event is only observed after a few minutes and on PEG no spinning is observed at all. In FOST-A-3 30% of the observed spores are spinning. Even if this number is significantly higher than for AWG (11%) the amount of spinning spores is decreasing from FOTS-A-1 (77% *hit and Stick*, 0% spinning), to FOTS-A-2 (34% hit and stick + 30% *spinning*) and FOTS-A-3 (0% *hit and stick*, 31% *spinning*). This decrease also correlates with the disappearance of the *hit and stick* pattern. Anyhow the first permanent settlement is not witnessed before FOTS-A-3 when the *hit and stick* pattern has completely vanished.

Furthermore, for all three subsets the *hit and run* pattern occurs only seldom (FOTS-A-1 (8%), FOTS-A-2 (6%), and FOTS-A-3 (2%)), meaning that only a few spores select the surface not to be suitable to settle after a brief contact, but most of

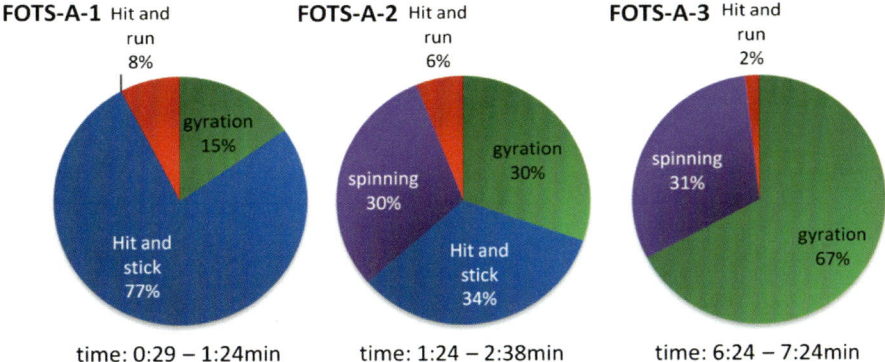

Fig. 6.5 Observed motion patterns on FOTS

the spores give the surface a try. This could be due to the attractiveness of the surface (spore–surface interaction), or it could be a collective effect caused by other spores spinning on the surface (spore–spore interaction).

In FOTS-A-3 the *gyration* pattern is the most dominant surface exploration pattern (67%). Its amount is similar to the amount of *gyration* on AWG (63%). Anyhow, at this point in time the exploration appears to be—based also on the other obtained parameters (e.g. enrichment, velocity distribution) discussed later— similar to the exploration on AWG.

To understand the observed high surfaces settlement in the standard AMBIO settlement assays the *hit and stick* pattern is the key even if it only occurs within the first 3 min after the spore injection into the wet cell.

In Appendix, Sect. A.3.1.1 the *hit and stick* pattern is discussed in detail. In Table A.8 all observed trajectories assigned to the latter are summarized. The unique feature of the *hit and stick* pattern is the *sticking* phase which leads to the observed surface trapping. On average the *sticking* phase last for 15 ± 12 s. The longest observed phase lasts 50 s. With elapsing time the duration of the *sticking* phase is getting shorter until the pattern vanishes completely. No spore trajectory not containing a *sticking* phase is assigned to the *hit and stick* pattern. In the reconstruction and the 3D analysis it can be depicted that the spore is not resting on the surface but rather try to break free from it. To draw a comparison the spore motion during the *sticking* phase seams similar to a bug lying on its back and trying to get back on its feet. Therefore the spores appear not to be "satisfied" with their position on the surface. The phase can be interpreted the way that the spores try to break free from the surface to reobtain their motility and to search for a more suitable place to settle.

Figure 6.6 shows the $\bar{a}_v$ distribution for the spores assigned to ASM (Active Searching Motion see Sect. 5.3) on AWG (panels (a)) and FOTS (panels (b–d)) at different points in time. In this figure the total number of spores is marked in blue whereas the subset swimming towards the surface is coloured red and the fraction swimming away from the surface is marked in green. The observed percentage of motion patterns are also shown in the figure.

In the water column (50–250 µm) for all experiments a similar trend is observed ($\bar{a}_v = 20 \pm 4°$). As also observed for PEG (shown and discussed in the last section) the value of $\bar{a}_v$ increases the closer a spore is swimming to the surface. The value of $\bar{a}_v$ starts to increase at a distance of 50 µm from the surface. During the *hit and stick* pattern the approach to the surface is unique. This can be seen in Fig. 6.6, panel (b). For the approaching spores (red curve) the value of $\bar{a}_v$ (30–40 µm $16 \pm 2°$, 20–30 µm $22 \pm 4°$, 10–20 µm $21 \pm 4°$, 0–10 µm $33 \pm 1°$) does not change considerably in comparison to the observed value for the bulk. This means that the spores do not perform more turns on the approach to the surface. In contrast at a later point in time on FOTS (FOST-A-3, panel (d)) the values of the $\bar{a}_v$ (30–40 µm $25 \pm 2°$, 20–30 µm $28 \pm 2°$, 10–20 µm, $34 \pm 2°$, 0–10 µm $53 \pm 2°$) increase significantly and are similar to the observed values on AWG (panel (a)). This increase can be correlated to the enhanced occurrence of the *gyration* pattern and the movement along the surface. Furthermore, in Fig. 6.6,

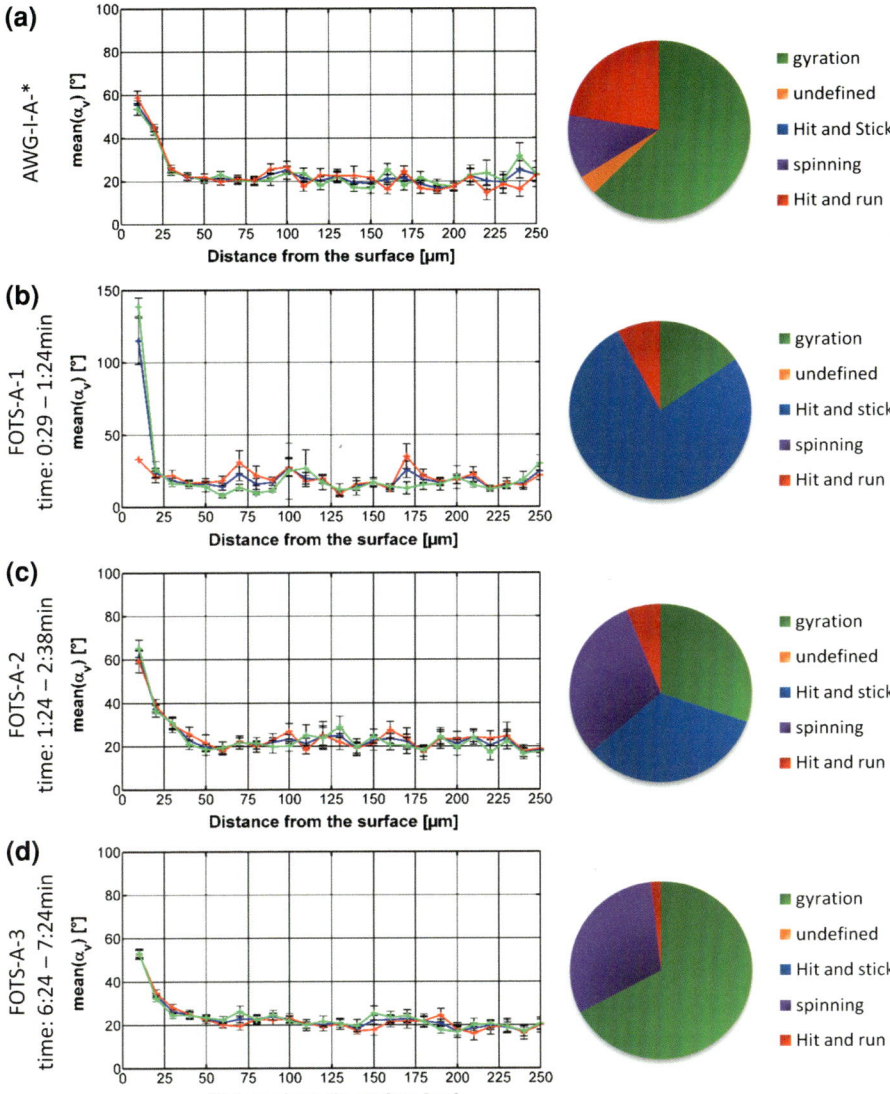

Fig. 6.6 Comparison of $\bar{a}_v$ and the observed motion patterns for **a** AWG, **b** FOTS-A-1 (please note the different scale on the y-axis), **c** FOTS-A-2, **d** FOTS-A-3. To study the approach towards and the detachment from the surface only spores assigned to the *gyration, hit and run* pattern and the approach/detachment part of the *hit and stick* pattern (abbreviated as ASM) are shown. The *sticking* phase and spores assigned to the *spinning* pattern are excluded, because their movement obtains no information for the approach towards and the detachment from the surface. In **a–d** the line color encodes the following: *red line*, towards the surface; *blue line*, all spores; and *green line*, away from the surface

panel (b) the value of $\bar{a}_v$ for the spores swimming away from the surface (green) is extremely high. Only the spores assigned to the *gyration* and *hit and run* pattern contribute to this peak, because the spores belonging to the *hit and stick* pattern stick to the interface and subsequently swim according to the *spinning* pattern and therefore do not leave the surface during the duration (54 s) of the experiment (FOTS-A-1). The high value of $\bar{a}_v = 139 \pm 6°$ for the spores swimming away at this point in time indicates that the spores perform sharp turns in the swimming direction and subsequently swim away from the surface. With the disappearance of the *hit and stick* pattern the surface exploration behaviour is getting similar to the observed motility on AWG (see Fig. 6.6, panels (a, c, d)).

The observation of the change in swimming direction ($\bar{a}_v$) can be connected with the flow towards the surface ($\bar{a}_z$) and swimming velocity (v_m) shown in Fig. 6.7. In this figure the same colour systematic is used as Fig. 6.6. For PEG (discussed in the last Sect. 6.2) and AWG (Fig. 6.7, panel (a)) for the complete volume (aside of the distances 0–20 µm from the surface) no preference to swim either towards or away from the surface is observed. This is also found for FOTS-A-2 (panel (c)) and FOTS-A-3 (panel (d)), as the $\bar{a}_z$ distribution for all recorded spores (blue curve) scatters within the error margin around 90°. No trend being observed for spores to swim towards or away from the surface can be correlated to enhanced occurrence of the *gyration* pattern. In contrast, during the time span the *hit and stick* pattern is the most dominant surface pattern (FOTS-A-1, panel (b), blue curve) the value of the $\bar{a}_z$ distribution is larger than 100° nearly throughout the complete volume. This means that the observed spores prefer to swim towards the surface.

The spores assigned to ASM (spores assigned: to *hit and run*, *gyration*, and the approach/detachment part of the trajectories belonging to the *hit and stick* pattern) and shown in Fig. 6.7 (FOTS-A-1, panel (b), red curve) approach the surface on average under a value of $\bar{a}_z = 100 \pm 1°$ (0–10 µm). In Appendix, Sect. A.3.1.1 the average approach angel is determined exclusively for the approach part of the trajectories assigned to the *hit and stick* pattern. Here the value of $\bar{a}_z$ is $140 \pm 26°$ which is considerably larger than the value for the spores belong to ASM ($\bar{a}_z = 100 \pm 1°$). Although the majority of the spores belong to the hit and stick pattern (77%, approach angle $\bar{a}_z = 140 \pm 26°$) the general approach angle ($\bar{a}_z = 100 \pm 1°$) is strongly influenced by the *gyration* and *hit and run* pattern. This decrease in the value of $\bar{a}_z$ is due to the fact that the spores assigned to the *gyration* pattern, do swim parallel the surface, leading to many low angles values in this regime and therefore an over representation of these trajectories.

Based on the mean velocity (v_m) distribution the surface interaction strength during the exploration behavior can be studied. The stronger the drop in the swimming speed the stronger the interaction. For FOTS-A-3 (37% decrease) and AWG (38% decrease) the slowdown is similarly pronounced (panels (e, h)). For both experiments the *gyration* pattern is the most dominant exploration pattern. The decline gradually starts at a distance of at least 50 µm from the surface and gets considerably steeper at a distance of 10–20 µm from the surface. The spores swimming within these sections are already able to touch the surface with the

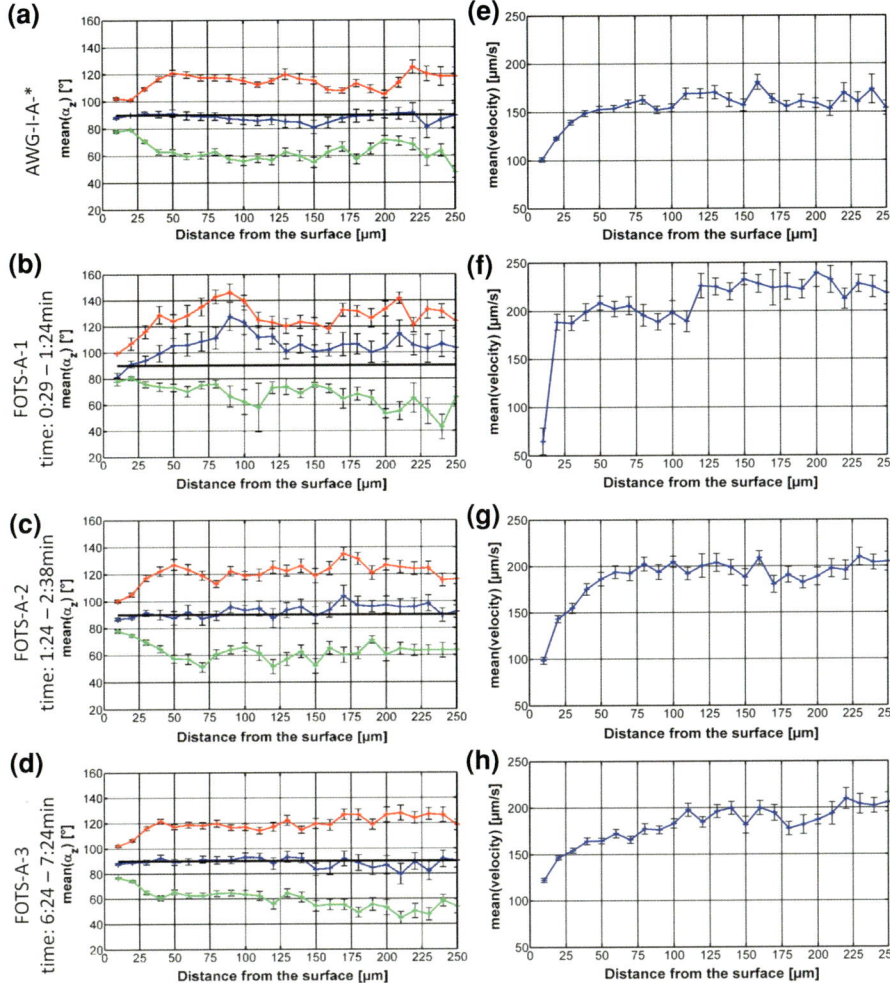

Fig. 6.7 Comparison of FOTS (**b–d**, **f–h**) versus AWG (**a**, **e**) for $\bar{a}_z$ (**a–d**) and v_m (**e–h**) distribution. In all panels the line color encodes the following: *red line*, towards the surface; *blue line*, all spores; and *green line*, away from the surface

flagella and therefore the slowdown represents the surface interactions. In FOTS-A-1 (panel (f)) during the *hit and stick* pattern the v_m distribution is notably different. The spores swim towards the surface with a constant velocity, nearly as fast as they swim in the bulk (aside of the drop in velocity around 110 µm). Only in the section 0–10 µm from the surface an extremely sharp drop in the velocity is observed. This slowdown is due to the strong spore–surface interactions leading to the observed *sticking* phase during the *hit and stick* pattern. For the trajectories assigned to the *hit and stick* pattern in Appendix, Sect. A.3.1.1 it is described that the velocity drops sharply from an average 200 µm/s towards zero. This change in velocity occurs at a distance of 20 ± 9 μm from the surface. Since, the spores approach the surface under an angle of $\bar{a}_z = 140 \pm 26°$ and the flagella are hold behind the body while swimming, the spore swims against the surface without havening contact between the flagella and the surface before. That the drop is observed at a distance of 20 ± 9 μm from the surface is due to the fact that the spore swims ≈ 20 μm between to consecutives frames and therefore the spore is observed that last time moving 20 μm from the surface.

In conclusion: the $\bar{a}_v$, $\bar{a}_z$ and v_m distributions change when the *hit and stick* pattern occurs. Based on this analysis it is possible to conclude that the spores "crash" into the surface and are subsequently trapped at the interface during the *hit and stick* pattern. For a spore trapped a *sticking* phase is observed until it is able to spin and subsequently leave the surface.

After the motility on FOTS explained in detail the questions: "Why does the *hit and stick* pattern occurs?" and "Why does it vanish already after approximately 3 min?" are still open and need to be further discussed. The next paragraph gives a possible explanation based on the literature but the phenomenon is not yet completely understood and some further experiments are necessary, because the focus of this thesis was to establish the foundations to determine and define 3D motion patterns.

A possible explanation for the *sticking* phase in the *hit and stick* pattern can be found by making a comparison to the observed anomalous spore settlement on cationic oligopeptide surfaces [13, 14]. On these surfaces some spores are attached sidewise and are not able to continue in their lifecycle. This sidewise attachment is explained by specific surface spore membrane interactions which trap the spore in an odd orientation on the surface [13]. Being held in this configuration the spore is not able to spin and continue in its life cycle. Nevertheless, most spores which are initially attached sidewise are strong enough to reorient themselves on the coating and either swim away or settle normally.

Since the behavior on FOTS appears similar to the one on the cationic oligopeptide surfaces, it can be interpreted similarly. Due to interactions with the surface the spores are initially stuck sideways or in an odd orientation to the surface. Subsequently the spore wriggles until it is reoriented. Following this reorientation, the spore always undergoes a *spinning* phase, even if the phase only lasts for a few seconds. If the spore leaves the surface, it is out of the *spinning* motion but not directly out of the *sticking* phase. It seems that only a small percentage of spores are capable of immediately leaving the surface, but rather spin extensively. This is the reason why so many *spinning* spores are observed on FOTS.

All spores on the FOTS coating are able to reorient themselves whereas on cationic oligopeptide surface it is only a subfraction of the spores. Therefore, on the cationic oligopeptide but not on the FOTS surface the anomalous spore settlement is observed in the standard settlement assay lasting 45 min. This difference has to be due to different forces holding the spores on the surface. The cationic oligopeptides are hydrophilic (contact angle (CA) = 36–45°) and the authors claim that the spores are held to the surface by specific interactions caused by the surface bound peptides and the spore membrane [13, 14]. Such specific interactions cannot cause the *sticking* phase on FOTS. There must be another reason. The fluorinated monolayer is covalently bond to the surface and therefore the coating is stable at the experimental conditions [15]. It does not leak into the solution and does not swell [15]. FOTS is similar to AWG in respect to structure (topology, porosity, periodicity) and stiffness (both are hard surfaces, Young modulus for AWG $\approx$ 69 GPa [16]). But the wettability of FOTS is different in comparison to AWG. FOTS is hydrophobic (CA $=$ 110.5°) whereas AWG is hydrophilic (CA $=$ 30°). Therefore the *hit and stick* pattern is probably caused by the hydrophobicity of the coating [17]. Whether the *hit and stick* pattern also occurs on a soft hydrophobic coating—like the commercial available Intersleek® coating of International Paint—has not been studied yet.

That the pattern vanishes could have several reasons and is need to proof by further experiments:

- The FOTS surface could rearrange during the immersion appears to be unlikely because the SAM is chemically bond to the surface and is composed out of a short ($C_8H_4F_{13}$) nearly completely fluorinated carbon chain.
- The surface is being "conditioned". There could at least three sources for the condition the surface:

  - Substances from the ASW
  - Polysaccharide/protein from the *Ulva* thalus or bacteria released in the ASW
  - The spores could "self condition" the surface by secretion of pre-glue, glue and signaling molecules during the *hit and stick* pattern.

The change in the behavior is most likely due to the fact that a conditioning film develops on the coating. This film reduces the hydrophobic forces and allows the spores to reorient themselves. The evidence that polysaccharides/proteins from the *Ulva* thalus or bacteria released in the ASW form a film on hydrophobic surfaces is based on not yet published results observed by Thome et al. (I. Thome, M.E. Pettitt, M.E. Callow, J.A. Callow, M. Grunze, A. Rosenhahn, in preaperation). In this study the development of a conditioning film on a hydrophobic surface (HS–$(CH_2)_{11}$–CH_3 SAM) was observed by ellipsometry. Additional to these substances the spore can "self condition" the surface. To distinguish between these two potential sources of conditioning is not possible without further experiment, but based on the referred study above (Thome et al. (I. Thome, M.E. Pettitt, M.E. Callow, J.A. Callow, M. Grunze, A. Rosenhahn, in preaperation)) the substances out of the ASW can be neglected for the formation of the condition layer. Whether a condition film also develops on hydrophilic surfaces is still under investigation. The existence of this film can explain why the spores are able to reorient themselves successfully on FOTS and why the *hit and stick* pattern vanishes.

Even if the *hit and stick* pattern last only for a short time it has further influence on the surface exploration, because some spores are always observed to be spinning on the surface. Therefore already in FOTS-A-3 the first settlement is witnessed. How this early settlement event influences ongoing settlement is not studied within the scope of this thesis, but the general spore settlement observed by the standard AMBIO assay can provide information. On hydrophobic surfaces in general a high amount of settlement is observed after a 45 min assay [18, 19]. Furthermore, the settlement is always observed to be in patches and is not consisting of evenly distributed individually settled spores [18, 19]. In these studies the authors could demonstrate that on a smooth surface the spores have an advantage to settle in groups, because they are better protected against shear stress and can adhere stronger to the surface. The earlier, e.g. forced by the trapping phase, the first settlement occurs the higher is the probability that another spore finds the already settled spore on the surface and settles next to it. With elapsing time this behavior leads to the formation of a spore cluster. It is known from literature that spores prefer to settle in depressions to be sheltered from shear stress [20]. On a smooth surface a solitarily settled spore can provide shelter for further organisms in its vicinity. Whether the spores are attracted by settled spores (quorum sensing) or whether they just find a settled spore randomly is not clear. However, an early settlement event induced by the trapping phase on FOTS could explain the high amount and patchy settlement on this surface. Schilp et al. [12] explained the raft formation on EG_6OH, where the adhesion strength is nearly zero, by an involuntary gliding of spores on the surface, after they have committed themselves to settlement and lost their ability to swim (a motion comparable to the diffusion of a physical adsorbed gas particle on a surface). Based on the results above this theory appears unlikely to explain the patch formation on hydrophobic coatings where the spores adhere strong enough to withstand a shear stress as high as 56 Pa [18]. An active selection of the spores to form patches appears to be a more likely explanation on hydrophobic surfaces.

Based on the results a hypothesis to explain the high amount and patchy spore settlement on hydrophobic coatings observed after 45 min in the AMBIO settlement assay can be proposed:

• A trapping phase in the beginning exists on all hydrophobic coatings.
• This short phase induces spinning of spores which leads to early settlement.
• After a conditioning film has formed and the *hit and stick* pattern has vanished on the FOTS surface the observed exploration behavior is similar to the behavior observed on AWG.
• An early settled spore can act as a nucleus for further spore settlement to form a cluster.

The nucleation hypothesis has to be verified by further 3D holographic motion analysis where the spatial spore density distribution of spores exploring the surface is studied in dependency of settled spores. The first experiment to verify this hypothesis has already been carried out but is not published yet. (I. Thome, M.E. Pettitt, M.E. Callow, J.A. Callow, M. Grunze, A. Rosenhahn, in preaperation) examined the spore settlement on hydrophobic surfaces in dependence of conditioning layer formation on the surface. In this case a conditioning layer is defined as a film which has formed out of molecules secreted by swimming spores. Comparing the surfaces with and without conditioning films the amount of settled spores and their appearance is completely different. Settlement on the hydrophobic coating is, as always, high and patchy, but on surfaces where assay a conditioning layer was formed prior to the settlement it is significantly smaller and the spores settle independently. Since it is shown in the 3D motion analysis that the *hit and stick* pattern does not occur if the conditioning film is formed on FOTS, it is very likely that on conditioned surfaces the *hit and stick* pattern will not occur. Therefore this observation verifies the theory that the trapping phase is important for the patchy and high amount of settlement and that individually settled spores on smooth surfaces can act as a nucleus for further and patchy settlement.

6.4 Hydrodynamic Trapping or Active Extended Exploration Near the Surface?

Within the motility study it is observed that the spores accumulate next to the surface. This enrichment extends up to a distance of 200 μm from the surface and exceeds the enrichment observed for other microorganisms. In literature for *E. coli* [21] and bull spermatozoa [22] an enrichment is observed up to a distance of 40 μm and is explained by hydrodynamic forces. The motility of *Ulva* spores can be classified in three zones.

(1) 0–50 μm from the surface
(2) 50–200 μm from the surface
(3) 200–∞ μm from the surface

The zone 3 (200–∞ μm from the surface) describes the motility in solution and is already discussed in Sect. 5.1. In the following the motility in zones 1 and 2 is discussed for the observed exploration behavior on the different surfaces and additionally in dependency to the change in the exploration behavior on FOTS.

Figure 6.8 shows the spore enrichment close to the investigated surfaces for the spores assigned to ASM. Furthermore, for a comparison the surface enrichment of *E. coli* [21] (red curve) and bull spermatozoa [22] (green curve) over a glass surface is shown in each panel. The distribution for these two microorganisms is not studied further than 100 μm from the surface so that the curve stops at this distance. The enrichment for each organism is normalized to the highest cell density.

If the *hit and stick* pattern does not occur (AWG-I-A-*, PEG-A-*, FOTS-A-3) the spores accumulate up to a distance of 200 μm from the surface (Fig. 6.8, panels (a, b, c)). For all these experiments the spore enrichment near the surface is of the same order and starts to increase at a distance of 200 μm from the surface. In this general trend further details are observed and can be explained if the swimming pattern distribution discussed before is envisioned (see Fig. 6.1). For AWG-I-A-* (panel (a), 63% *gyration*, 22% *hit and run*) and FOTS-A-3 (panel (d), 67% *gyration*, 2% *hit and run*) the *gyration* pattern occurs by a similar percentage. In contrast, on PEG (panel (b), 57% *gyration* and 42% *hit and run*) the *gyration*

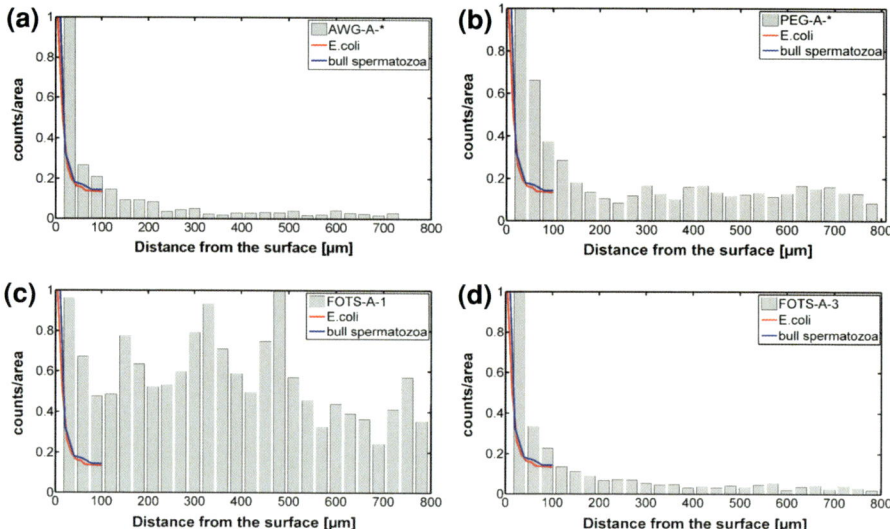

Fig. 6.8 Spore enrichment near the investigated surfaces and in comparison to the enrichment of Bull spermatozoa [22] and *E. coli* [21]. **a** AWG-A-*; **b** PEG-A-*; **c** FOTS-A-1; **d** FOTS-A-3

pattern is slightly less populated, but more importantly, the *hit and run* pattern occurs with a high percentage, meaning that the spores do not stay at the interphase but rather swim away. This is also visible in the spore distribution shown in panel (b) because the observed spore concentration at the section 30–60 µm (second bin in the histogram) is considerably larger for PEG (panel (b)) than for FOTS-A-3 (panel (d)) or AWG-I-A-* (panel (a)). Furthermore, on PEG the ratio of the amount of spores near (0–200 µm) the surface and the amount of spores in the bulk (200–800 µm) is smaller ($1/0.13 = 7.7$) as in comparison to AWG (panel (a), $1/0.03 = 33.4$) and FOTS-A-3 (panel (d), $1/0.03 = 33.4$). Nevertheless for AWG-I-A-*, PEG-A-* and FOTS-A-3 the increase in the spore accumulation is observed to start at a distance of 200 µm from the surface.

In contrast the described spore enrichment is not observed when the *hit and stick* pattern occurs (FOTS-A-1, Fig. 6.8, panel (c)). During the *hit and stick* pattern, as discussed in detail in the last Sect. 6.3 the spores are trapped at the interface. On AWG-I-A-*, PEG-A-* and FOTS-A-3 the spore accumulation is caused by the motility of the spores belonging to the *gyration* and *hit and run* patterns. Spores assigned to these patterns are only present in a small amount during experiment FOTS-A-1. Therefore no spore enrichment can build up close to the surface.

In Fig. 6.8 the enrichment of *E. coli* (red curve) and bull spermatozoa (blue curve) near a surface is shown for comparison [21, 22]. For both organisms an extremely similar distribution is observed and the concentration increases at a distance of 40 µm from the surface. The accumulation in vicinity to the surface is significantly smaller for *E. coli* and bull spermatozoa than for *Ulva* spores. *Ulva*

Table 6.1 Size, shape, swimming speed and Re for *E. coli*, *Ulva* spores and bull spermatozoa

Organism	Size	Shape	Swimming speed	Re
E. coli	0.5 × 1 μm [23]	Bar-shaped	≈ 30 μm/s [23]	3×10^{-5}
Ulva spores	2.5 × 2.5 μm	Round	≈ 150 μm/s	4×10^{-4}
Bull spermatozoa	10 × 5 μm [24]	Cudgel-shaped	≈ 100 μm/s average trajectory speed for circular swimmers [25]	7×10^{-4}

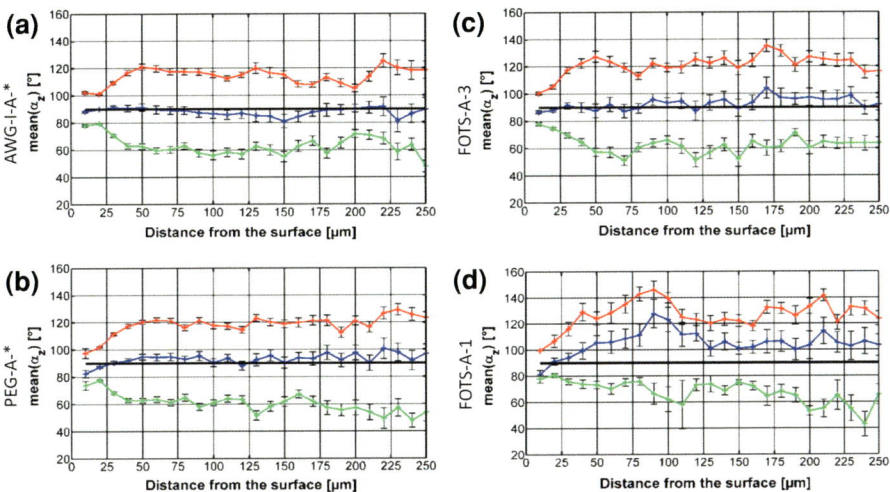

Fig. 6.9   Anisotropy of spore movement for AWG (**a**), PEG (**b**), FOTS-A-3 (**c**), and FOTS-A-1 (**d**)

spores, *E. coli*, and bull spermatozoa have different sizes, swimming speeds, and shapes (summarized in Table 6.1). Therefore they swim in different Reynolds number regime (Re). For *E. coli* the Re is 3×10^{-5} whereas for bull spermatozoa the Re is 7×10^{-4}. Nevertheless the accumulation in vicinity to the surface is for of *E. coli* and bull spermatozoa nearly identical. For these two organisms the enrichment next to the surface is explained by hydrodynamic forces trapping the organisms at the interface [21, 22] (see also Sect. 3.3.3).

To understand the response to hydrodynamic forces on the motility of *Ulva* spore the change in the $\bar{a}_z$ distribution can provide further details. In Fig. 6.9 the $\bar{a}_z$ distribution is shown for AWG-I-A-*, PEG-A-* and FOTS-A-3 (panels (a–c)). All curves show a similar trend.

As explained in detail in Sect. 4.4 (Fig. 4.17) the $\bar{a}_z$ distribution is isotropic if its mean value for the complete angle range is 90° (marked blue) and the value of $\bar{a}_z$ for the spore fraction swimming towards the surface is 124° (marked red) and 56° for the fraction swimming away (marked green).

In Fig. 6.9 the $\bar{a}_z$ distribution (blue curve) for AWG-I-A-* (panel (a)), PEG-A-* (panel (b)) and FOTS-A-3 (panel (c)) has a value of ≈ 90° within the error margin for the complete observation volume (taking aside the first data point). This means that no trend is observed for the spores to swim either towards or away from the surface. Nevertheless the movement is not isotropic for the complete observation volume. Close to the surface it is observed that the spores swim along the surface whereas in the water column the spores move in a random orientation. This direction of movement can be studied by the $\bar{a}_z$ distribution of the spore fraction swimming towards (red curve) or away (green curve) from the surface. From deep in the water column up to a distance of 50 μm from the surface the observed $\bar{a}_z$ value for the spore fraction swimming towards (red curve) and away (green curve) from the surface is close to the expected mean values of an isotropic distribution for AWG-I-A-* (panel (a)), PEG-A-* (panel (b)) and FOTS-A-3 (panel (c)). For the spore fraction swimming towards the surface a value of $\bar{a}_z \approx 124°$ and for the fraction swimming away a value of $\bar{a}_z \approx 56°$ is observed. Closer than 50 μm from the surface the $\bar{a}_z$ distribution for the spore fraction swimming towards the surface as well as for the fraction swimming away from the surface changes towards 90°. This change in the $\bar{a}_z$ distribution for both spore fractions means that the spores change their swimming direction and start to swim parallel to the surface. The change is symmetrical for both spore fractions and decreases up to the section 10–20 μm from the surface. For the section 0–10 μm the $\bar{a}_z$ value does not further decrease and remains the same. Already in the section 10–20 μm the spores can touch the surface with their flagella while swimming and therefore the distribution in these two sections is similar.

The change from an isotropic motion in solution to a motion along the surface is caused by hydrodynamic forces. This can be verified by the fact that the change occurs at the distance from the surface where the hydrodynamic forces increase and (even more important) that the change in the swimming direction is symmetrical for swimming towards and away from the surface. While swimming the spore is surrounded by a flow field. This flow field is altered by the hydrodynamic forces forcing the spore to change its swimming direction. Berke et al. [21]. and Lauga and Powers [26]. predicted in a theoretical model that hydrodynamic forces near a surface cause *E. coli* to swim parallel to it. Furthermore they calculated that while approaching a surface the hydrodynamic forces gradually force the organisms to change their direction of motion to a direction parallel to the surface. By the motility analysis of *Ulva* spores we measured this hydrodynamic forced change in the swimming direction for a microorganism, to our knowledge, for the first time in detail.

In summary, in zone 1 (0–50 μm) the hydrodynamic forces near a solid interface force a swimming spore to change its direction of movement and swim along the surface. This change starts at a distance of 50 μm from the surface. However, the spore enrichment is observed up to a distance of 200 μm from the surface and can therefore not be caused by hydrodynamic surface trapping as observed for *E. coli* and spermatozoa. Rather it has to be due to the exploration behavior of spores. To characterize the spore enrichment in vicinity to the surface the obtained histograms presented in Fig. 6.8 are fitted with an exponential

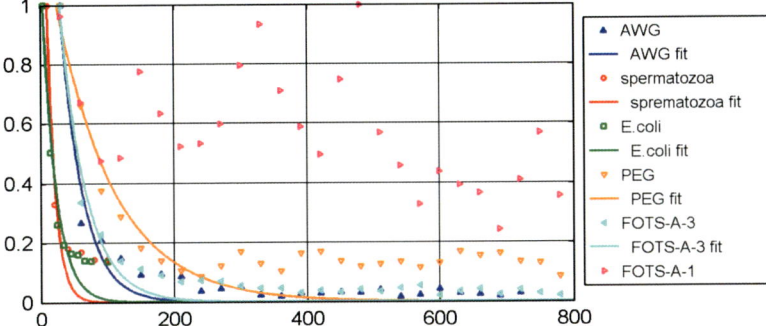

Fig. 6.10 Spore enrichment near the investigated surfaces shown already in Fig. 6.8 as a bar graph and fitted by an exponential function (see Eq. 6.1)

function (Eq. 6.1) leaving the free parameters a and b. The result is shown in Fig. 6.10.

$$f(x) = a \cdot \exp(-bx) \tag{6.1}$$

By the fit three different surfaces enrichment distances can be observed. For *E. coli* and bull spermatozoa the cell enrichment near the surfaces increases at a distance of 40 μm from the surface. For FOTS-A-3 and AWG-A-* the accumulation extends up to 200 μm. The observed exploration behavior on FOTS-A-3 and AWG-A-* is very similar after the *hit and stick* pattern has vanished (see discussion in Sect. 6.3, similar decrease in the velocity close to the surface, similar percentage of *gyration* pattern). On the deterrent PEG coating (see discussion 6.3, high percentage of *hit and run* pattern, no *spinning* pattern, depletion layer) the spore enrichment exists further into the solution than observed for the other surfaces. This means that the spore motility alters the spore accumulation near the surface of dependence to the explored surface. The results of the spore enrichment near a surface provide evidence that the spores react to the surface at a greater distance as the reach of the hydrodynamic forces (0–50 μm) which force the spore to swim parallel to the surface.

Figure 6.11, panels (c, d, f, h) show the mean velocity (v_m) distribution for all investigated surfaces from the surface up to 800 μm into the water column. The section 0–200 μm is highlighted and magnified in the additional panels (a, c, e, g). In general (aside from FOTS-A-1 panels (g, h), time period where the *hit and stick* pattern occurs) the spores swim slower the closer they get to the surface. The distance from the surface where the swimming speed starts to slow down is in consistency with the distance where the spore concentration increases. What causes the spores to slow down is not yet understood but it has to be related to the presence of the surface. Apart from FOTS-A-1 the slowdown in the swimming speed is pronounced in the section 0–50 μm from the surface. A possible explanation for the large extended spore enrichment into the solution could be that while searching for a suitable settlement place the spores outswim the boundary layer

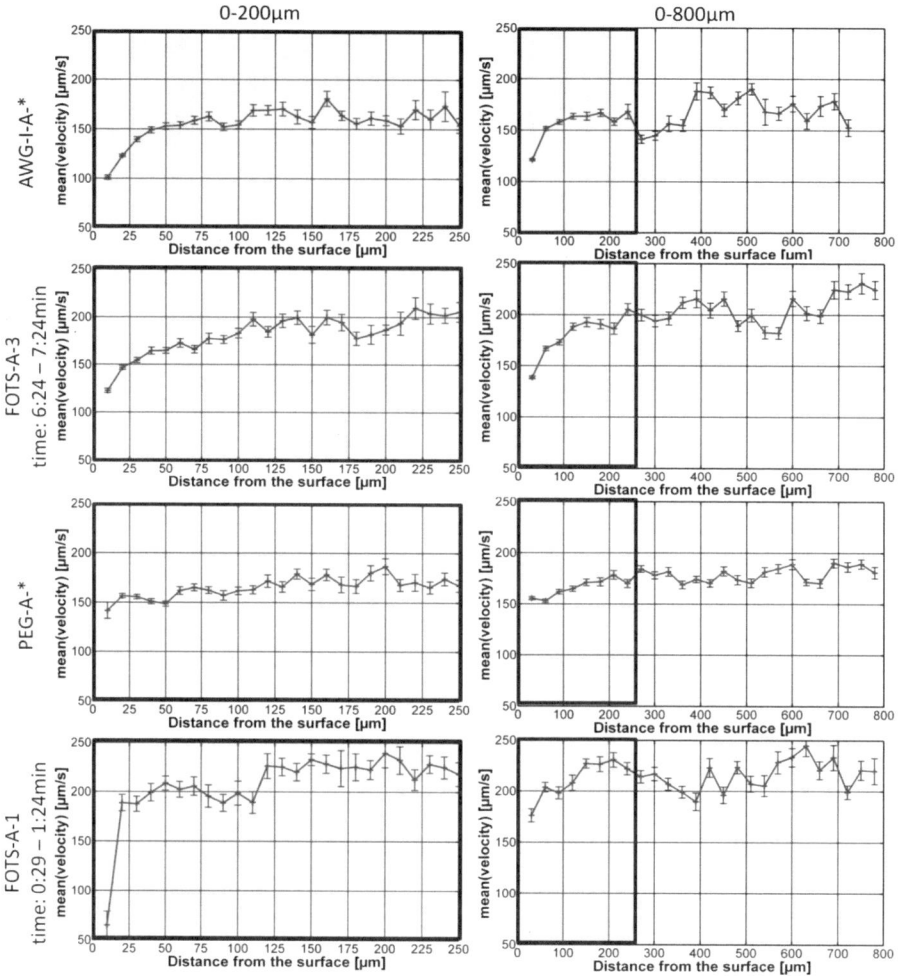

Fig. 6.11 Velocity distribution (v_m) for the investigated surfaces. **a, c, e, g** the v_m distribution for the section 0–200 μm; **b, d, f, h** the v_m distribution for the section 0–800 μm. The highlighted parts in **a, c, e, g** are magnifications of the highlighted parts of the distributions shown in **b, d, f, h. a, b** AWG-A-*; **c, d** FOTS-A-3; **e, f** PEG-A-* and **g, h** FOTS-A-1

(0–50 μm) in which they are slowed down before they start a new surface approach. This postulated spore exploration pathway could insure that the spores can effectively "sense" a large surface area. However, during this movement the spore does not move further away from the surface than it still can detect the surface, otherwise the exploration behavior would be ineffective. Which mechanism the spore uses to detect the surface is not known. To proof this explanation is the challenge for future work.

During the occurrence of the *hit and stick* pattern (FOTS-A-1, Fig. 6.9, panel (d)) detailed information of the spore exploration can be obtained. This is due to the fact that during the occurrence of *hit and stick* the spores are trapped at the surface and therefore the approach towards the surface can be studied without the "backflow" from the surface. It is not possible to follow this behavior the *hit and stick* pattern has vanished or on the other surfaces where this pattern does not occur. Furthermore, the motility analysis of FOTS-A-1 is started 30 s after the spores' injection. Therefore, the arriving spores at the surface have most probably found an interface for the first time. This argumentation is also valid for the other surfaces. Still on the surfaces without *hit and stick* pattern the first spores arriving at the surface explore the surface and swim according to the *gyration* pattern, meaning swimming towards and away from the surface. Since the spores accumulate near the surface and can enter the observation volume from all sides (aside from the surface side) it is not possible to discriminate whether a spore is in the area where it can feel the surface (e.g. by hydrodynamic forces) for the first time or not. Therefore the initial spore approach to the surface can only be studied during the *hit and stick* pattern.

In Fig. 6.9, panel (d) the $\bar{a}_z$ distribution is shown for FOTS-A-1. The value of $\bar{a}_z$ (blue curve) is bigger than $100°$ nearly for the complete observation volume. This means that the spores prefer to swim towards the surface even from a very large distance. The flow towards the surface is connected to the fact that the recording is started only 30 s after the spore injection into the wet. The witnessed spores are the first spores which enter the field of view. To obtain the observed spore enrichment in vicinity to the surface there must be a preference to swim towards the interface for a certain time span otherwise no enrichment can develop. If the enrichment has build up a balanced condition is achieved where as in the bulk no trend to swim towards the surface is observed anymore. On AWG and PEG spore enrichment close to the surface builds up in a very short time because spores explore the surface by swimming according to the *gyration* pattern. For FOTS-A-1 the spores are trapped at the interface and therefore the time until the enrichment (or the balance state) is reached takes until the *hit and stick* pattern does not occur anymore (FOTS-A-3). At this later point in time (FOTS-A-3) the value of the $\bar{a}_z$ distribution is comparable to the values for AWG and PEG (see Fig. 6.9) and no trend to swim towards the surface is observed. Based on these arguments and on the fact that the surface is stable under the experimental conditions it appears unlikely that swimming towards the surface observed for FOTS-A-1 is triggered chemotactically by molecules diffusing out of the coating into the water column.

The spore fraction approaching the surface (red curve, Fig. 6.9, panel (d)) provides further details of the spore motility. Aside from some outliers the spores approach the surface under a mean value of $\bar{a}_v \approx 124°$ until a distance of 110 µm from the surface. At a distance of 90 µm from the surface a peak ($146 \pm 6°$) is observed, which can be interpreted in the way that the spores actively change their direction into a steeper approach path to the surface. Based on the discussion above this distance is nearly twice as big as the distance at which hydrodynamic

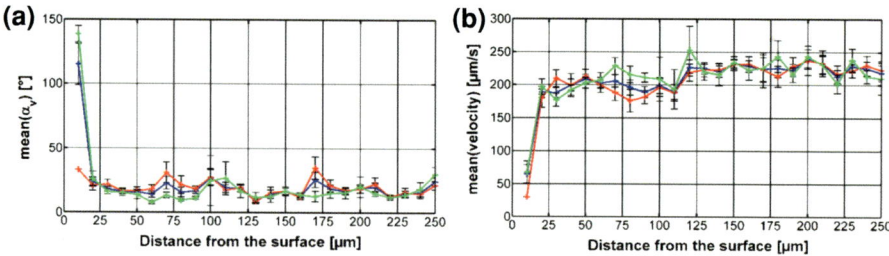

Fig. 6.12 Distribution of $\bar{a}_v$ (**a**) and v_m (**b**) for FOTS-A-1

forces start playing a dominant role on the spore motility. This change in swimming direction (start 110 µm, peak 90 µm from the surface) coincides with a drop in swimming velocity shown in Fig. 6.12, panel (b). In Fig. 5.7 it is shown that spores swim slower when turning. Therefore the concordant between the peak in the $\bar{a}_z$ distribution and the drop in the v_m distribution supports the interpretation that the spores actively turn and swim towards the surface. In the $\bar{a}_v$ distribution this change is identifiable but not strongly pronounced. However, why this peak only occurs on FOTS-A-1 and how it is connected to the occurrence of the *hit and stick* pattern is not yet completely understood and needs to be confirmed by further experiments. As discussed in detail in Sect. 6.3, during the *hit and stick* pattern the spores approach the surface in a unique way which differs significantly from the motility observed for all other investigated surfaces. A possible answer could be that this behavior stems from spores approaching the area in which they can detect the surface for the first time, and therefore change their approach direction into a steeper path straight towards the surface.

In conclusion: *Ulva* spores are, unlike *E. coli* and spermatozoa, not trapped by hydrodynamic forces near surfaces. Nevertheless during the approach to the surface, the hydrodynamic forces near (0–50 µm) the surface force the swimming organisms to change their swimming direction to a motion parallel to the surface. For *Ulva* spores the change in the swimming direction is observed by the change of the $\bar{a}_z$ distribution. For *E. coli*, spermatozoa and *Ulva* spores the distance from the surface, where the hydrodynamics are strong enough to alter the swimming motion, is nearly identically. Nevertheless the spore enrichment is extended to a distance from the surface of four times of the hydrodynamic trapping distance. Within the area of the enrichment the spores are observed to swim slower the closer they get to the surface. Furthermore, only during occurrence of the *hit and stick* pattern (FOTS-A-1, no surface enrichment, no slow down) at a distance of 110 µm from the surface it is observed that the spores change their swimming direction into a stepper approach towards the surface. These results show that the spores are able to responds to the surface before the hydrodynamic forces near the surface alter the swimming direction. How the spores detect the surface is an unresolved question in this thesis and is open to speculations and efforts to design further experiments.

References

1. M. Heydt, A. Rosenhahn, M. Grunze, M. Pettitt, M.E. Callow, J.A. Callow, J. Adhes. **83**(5), 417–430 (2007)
2. I. Thome, Einfluss der Oberflächenchemie und -Morphologie auf die Besiedelung von Oberflächen durch maritime Organismen, Diploma Thesis, Ruprecht-Karls-University of Heidelberg, Heidelberg, 2009
3. M.E. Callow, J.A. Callow, J.D. Pickett-Heaps, R. Wetherbee, J. Phycol. **33**(6), 938–947 (1997)
4. J.A. Callow, M.E. Callow, *The Ulva Spore Adhesive System* (Springer, Berlin, 2006), pp. 63–78
5. J.A. Callow, M.E. Callow, Phycologia **44**(4), 35 (2005)
6. I. Inouye, T. Hori, Protoplasma **164**(1–3), 54–69 (1991)
7. S. Miyamura, Cytologia **69**(2), 197–201 (2004)
8. K.L. Prime, G.M. Whitesides, J. Am. Chem. Soc. **115**(23), 10714–10721 (1993)
9. P. Harder, M. Grunze, R. Dahint, G.M. Whitesides, P.E. Laibinis, J. Phys. Chem. B **102**(2), 426–436 (1998)
10. S. Schilp, A. Rosenhahn, M.E. Pettitt, J. Bowen, M.E. Callow, J.A. Callow, M. Grunze, Langmuir **25**(17), 10077–10082 (2009)
11. S.I. Jeon, J.H. Lee, J.D. Andrade, P.G. Degennes, J. Colloid Interface Sci. **142**(1), 149–158 (1991)
12. S. Schilp, A. Kueller, A. Rosenhahn, M. Grunze, M.E. Pettitt, M.E. Callow, J.A. Callow, Biointerphases **2**(4), 143–150 (2007)
13. T. Ederth, P. Nygren, M.E. Pettitt, M. Ostblom, C.X. Du, K. Broo, M.E. Callow, J. Callow, B. Liedberg, B. Liedberg, Biofouling **24**(4), 303–312 (2008)
14. T. Ederth, M.E. Pettitt, P. Nygren, C.X. Du, T. Ekblad, Y. Zhou, M. Falk, M.E. Callow, J.A. Callow, B. Liedberg, Langmuir **25**(16), 9375–9383 (2009)
15. X. Cao, Antifouling Properties of Smooth and Structured Polyelectrolyte Thin Films, Ph.D. Dissertation, Ruprecht-Karls-University of Heidelberg, Heidelberg, 2008
16. Wikipedia, Web Page, Young's modulus, http://en.wikipedia.org/wiki/Young%27s_modulus. Accessed: 16 Oct. 2009
17. B. Wigglesworth-Cooksey, H. van der Mei, H.J. Busscher, K.E. Cooksey, Colloids Surf. B Biointerphases **15**(1), 71–80 (1999)
18. J.A. Finlay, M.E. Callow, L.K. Ista, G.P. Lopez, J.A. Callow, Integr. Comp. Biol. **42**(6), 1116–1122 (2002)
19. M.E. Callow, J.A. Callow, L.K. Ista, S.E. Coleman, A.C. Nolasco, G.P. Lopez, Appl. Environ. Microbiol. **66**(8), 3249–3254 (2000)
20. M.E. Callow, A.R. Jennings, A.B. Brennan, C.E. Seegert, A. Gibson, L. Wilson, A. Feinberg, R. Baney, J.A. Callow, Biofouling **18**(3), 237–245 (2002)
21. A.P. Berke, L. Turner, H.C. Berg, E. Lauga, Phys. Rev. Lett. **101**(3), 038102(1)–038102(4) (2008)
22. L. Rothschild, Nature **198**(4886), 1221–1222 (1963)
23. E.M. Purcell, Am. J. Phys. **45**(1), 3–11 (1977)
24. C. Brennen, H. Winet, Annu. Rev. Fluid Mech. **9**, 339–398 (1977)
25. F.R. Hallett, T. Craig, J. Marsh, Biophys. J. **21**(3), 203–216 (1978)
26. E. Lauga, T.R. Powers, Rep. Prog. Phys. **72**(9), 36 (2009)

Chapter 7
Conclusion and Outlook

In this thesis, the development of a transportable in-line holographic microscope for visible light was achieved and successfully used to track the swimming path of *Ulva* zoospores in 3D for the first time. Furthermore, a software program was developed and programmed from scratch to allow a fast and accurate position determination [1]. In addition many different analysis tools were developed to examine and classify the exploration behavior of *Ulva* spores.

Due to the large focus depth of the holographic technique it was possible to examine the motility of microorganisms in a large wet cell with a depth of 5,000 µm. Therefore, for the first time, the motility in solution (in an area without any boundary effects of the surface) and the exploration in vicinity to surface was studied at the same time in the same experiment. The motility close to the surface is different to the observed motility in solution. In the course of the study two different kinds of spores are identified, one of which does not explore the surface. The increase of the hydrodynamic forces in the vicinity (0–50 µm) to the surface forces the spores to swim parallel to the surface which is predicted by theory [2] but, to our knowledge, measured in detail for the first time. Nevertheless, the spores accumulate from the surface up to a distance of at least 200 µm from the surface (or greater depending on the investigated surface). In this area the mean swimming velocity of the spores decreases. Based on these observations and on the change in the approach angle observed on FOTS in the beginning of the recording it is possible to conclude that the spores interact with the interface at distances of at least 100 µm from the surface.

To classify the spore motility the movement is analyzed and assigned to different motion patterns [3]. The motility close to the surface is analyzed in detail in dependence of the surface properties and is set into context to the fouling kinetic on these surfaces, determined by additional experiments. The spore motility provides inside views of the settlement behavior in relation to the surface. The different motility observed in the vicinity to the surfaces goes along with the observed fouling kinetics determined by settlement assays. Thereby, in contrast to

M. Heydt, *How Do Spores Select Where to Settle?*, Springer Theses,
DOI: 10.1007/978-3-642-17217-5_7, © Springer-Verlag Berlin Heidelberg 2011

the duration of the 45 min settlement assay, it is sufficient to study the motility for 2 min within the first 5 min of surface exploration to predict the fouling kinetics.

On the deterrent PEG coating it is observed that the spores stay away from the interface. The flagella-surface contact is sufficient to disturb the exploration behavior so that the spores swim away from the surface. On the FOTS coating the exploration behavior is time dependent which can most probably be linked to a development of a conditioning film on the surface which alters the surface properties. In the beginning the spores are trapped at the interface. This trapping phase induces spinning which leads to settlement. The early settled spores act as a nucleus for further spore settlement which leads to spore clusters on the surface. This mechanism could explain the fouling kinetics observed by the standard assays.

For future developments it is planned to develop completely automated analysis software for the position determination and analysis of trajectories. In the last couple of months promising results were achieved to determine the position of algae automatically [4]. This new software will provide further possibilities to study the motility of microorganisms in greater detail and for longer time periods because even more motion data is feasible to process. In reference to the results shown in this thesis and in cooperation with the Prof. B. Rosenhahn, University of Hannover, Germany we successfully applied for a DFG grant to accomplish this project.

Parallel to the software development further experiments with microorganisms are planned and, to some extent, were already carried out. Based on the results achieved in this thesis another grant from the Office of Naval Research (ONR) is successfully obtained to carry out these studies. In the near future it is planned to study different microorganisms, especially bacteria, with the developed setup and the implemented analysis software. To achieve the required resolution to track bacteria a new camera was already bought. Additionally it is planned to study the antifouling properties in a more realistic environment by deploying the instrument in the ocean. This work will be done together with Prof. G. Swain, Institute of Technology, Florida, USA. The work with *Ulva* spores will be continued to answer the open questions raised in course of the thesis. Furthermore, the exploration behavior will be studied on different types of surfaces. The cationic oligopeptides [5] mentioned in Sect. 6.3 offer a great possibility to study the spore approach towards the surface. A topographic interesting surface is the sharklet AF^{TM} pattern [6] on which an extreme low spore settlement is observed. First promising experiments have already been carried out to study the exploration behavior on this surface.

References

1. M. Heydt, P. Divós, M. Grunze, A. Rosenhahn, Eur. Phys. J. E. **30**, 141–148 (2009)
2. E. Lauga, T.R. Powers, Rep. Prog. Phys. **72**(9), 36 (2009)

3. M. Heydt, A. Rosenhahn, M. Grunze, M. Pettitt, M.E. Callow, J.A. Callow, J. Adhes. **83**(5), 417–430 (2007)
4. L. Leal Taixé, M. Heydt, A. Rosenhahn, B. Rosenhahn, Automatic tracking of swimming microorganisms in 4D digital in-line holography data. IEEE Workshop on Motion and Video Computing (WMVC), Snowbird, Utah, Dec 2009, p. 8
5. T. Ederth, M.E. Pettitt, P. Nygren, C.X. Du, T. Ekblad, Y. Zhou, M. Falk, M.E. Callow, J.A. Callow, B. Liedberg, Langmuir **25**(16), 9375–9383 (2009)
6. J.F. Schumacher, M.L. Carman, T.G. Estes, A.W. Feinberg, L.H. Wilson, M.E. Callow, J.A. Callow, J.A. Finlay, A.B. Brennan, Biofouling **23**(1), 55–62 (2007)

Appendix

A.1 Exploration Behavior on AWG

The following exploration pattern analysis is based on 178 individual traces and 14,807 data points. Table A.1 gives an overview of the available statistics. Two experiments carried out with spores harvested at different collection trips (trip A and B) are analyzed. The name of the experiments follows the same systematic as described before (see Chap. 5). The experiments are analyzed at different points in time to investigate the time dependence of the spore motility in the vicinity of AWG surface.

Figures A.1 and A.2 show the obtained motion data. Both figures illustrate the same data set. While Fig. A.1 shows the trajectories up to a distance of 700 μm from an overview, Fig. A.2 scales only up to a distance of 200 μm for a more detailed depiction of the surface exploration behavior. Similar to the observations for the bulk swimming pattern no swarm behavior is detected in the motion data close to the surface. The spores swim individually and independently. The reported erratic, random motion in solution is disturbed by the presence of the surface. The influence of the surface is best seen in the xz projection in Figs. A.1 and A.2, panels (k–o). With elapsing time the spore accumulation close to the surface increases. The differences in spore motility are more obvious when individual trajectories are analyzed. For the following discussion the focus will be on experiment AWG-I-A-* because it is analyzed at several different time points.

A.1.1 Exploration on AWG: Swimming Pattern Analysis

In the following the motility patterns are discussed for individual, exemplary traces and are divided in two subsections: *gyration* and *spinning*. To understand the presented results it is important to remember the spore physiology (see Sect. 3.1). The spore has a droplet-shaped body with a diameter of 5 μm and

Table A.1 Statistics for the exploration behavior on AWG

Name	Distance				Elapsed time (min:s)	Duration (s)
	0–50 µm		50–200 µm			
	Number of traces	Number of data points	Number of traces	Number of data points		
AWG-I-A-1	13	625	19	377	0:35	41.9
AWG-I-A-2	37	1,493	41	1,030	3:14	41.8
AWG-I-A-3	25	1,713	24	553	11:57	42.0
AWG-II-B-1	24	2,720	15	501	5:00	102.8
AWG-II-B-2	40	4,156	347	1,653	22:39	102.4
Sum	139	10,707	133	4,100		303.9

attached to the body four flagella with a length of 15 µm. This means that a spore is able to "touch" a surface physically from a distance of about 20 µm away from the surface.

The exemplary trajectories are named according to the following systematic: experiment name (AWG-I-, AWG-II), motion pattern abbreviation (Gy: *gyration*, H&R: *Hit and run*, Sp: *spinning*), continuously number, e.g., AWG-I-H and R-1. The occurrence of the determined motion pattern is shown in Table A.2. All shown trajectories are discussed with the help of five individual plots. In the corresponding Figs. (A.3, A.4, A.5, A.6, A.7, A.8, A.9, A.10, A.11 and A.12) panel (a) shows a 3D rendered plot. This plot gives an overview of the motion. Panel (b) shows a velocity histogram with a fitted Maxwell–Boltzmann distribution (see Eq. (5.1)). In panels (c–e) detailed exploration parameters are shown and plotted versus the observation time in seconds. These parameters are the following:

- The change in the Z-position is shown in panels (c, d) in blue and the corresponding scale on the right side of the graph.
- The change in velocity is shown in panels (c, e) in black with the scale on the left side of the graph.
- α_v distribution is plotted in red in panels (d, e). Depending whether it is plotted together with the change in the Z-position panel (d) it is shown on the left side, or if it is plotted together with the change in velocity panel (e) it is plotted on the right side.

A.1.1.1 AWG Swimming Pattern: Gyration

Many spores stay in vicinity to the surface once they have gotten close to it. The motion pattern itself is very variable and the most dominant motion pattern on AWG (see Table A.2). The *gyration* pattern is assigned to a general motion pattern and already defined in Sect. 5.2.4.3. The swimming pattern *gyration* is illustrated with three exemplary trajectories shown in Figs. A.3, A.4 and A.5.

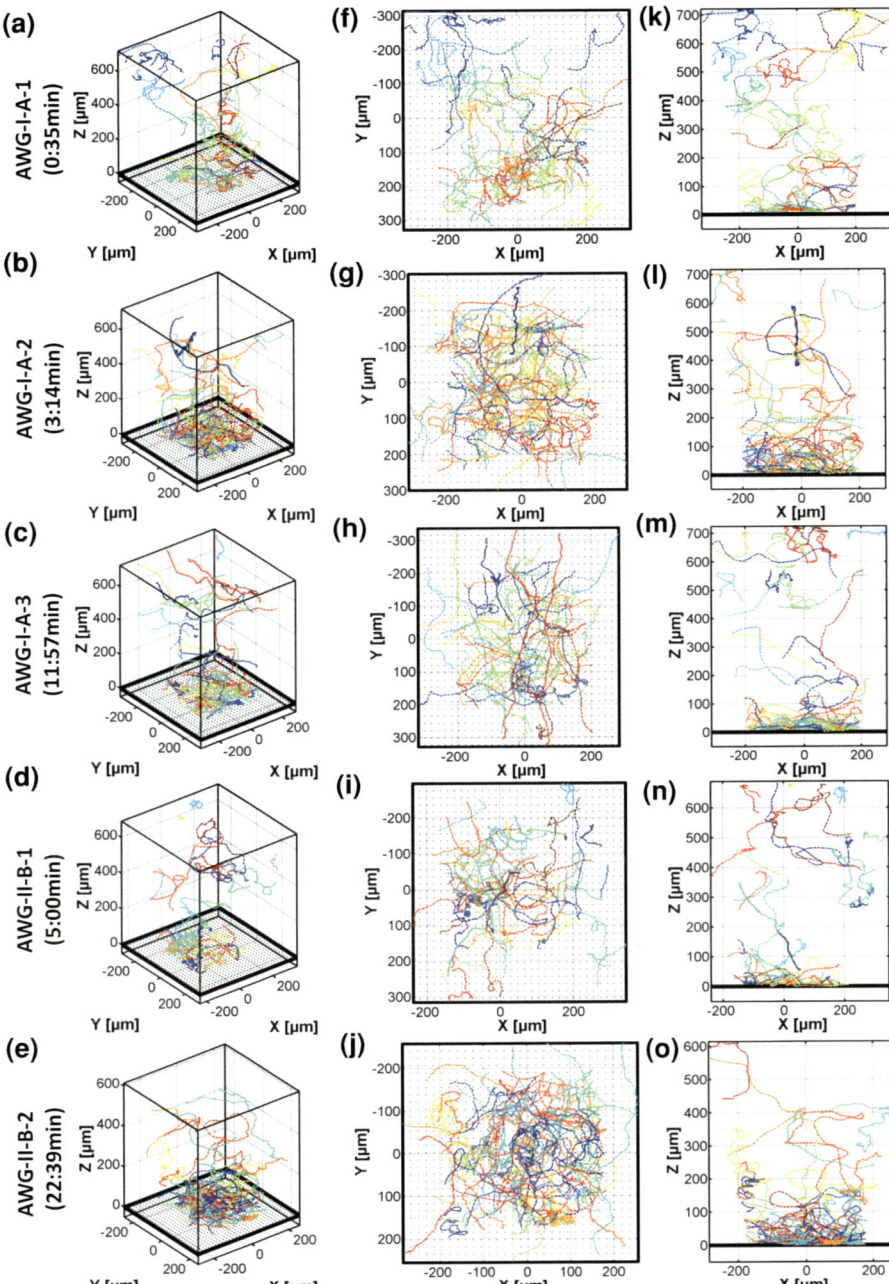

Fig. A.1 3D rendered plots of spores within a distance of 0–700 μm from the surface; **a–e** 3D view; **f–j** xy view; **k–o** xz view. To distinguish between trajectories they are marked in different colors

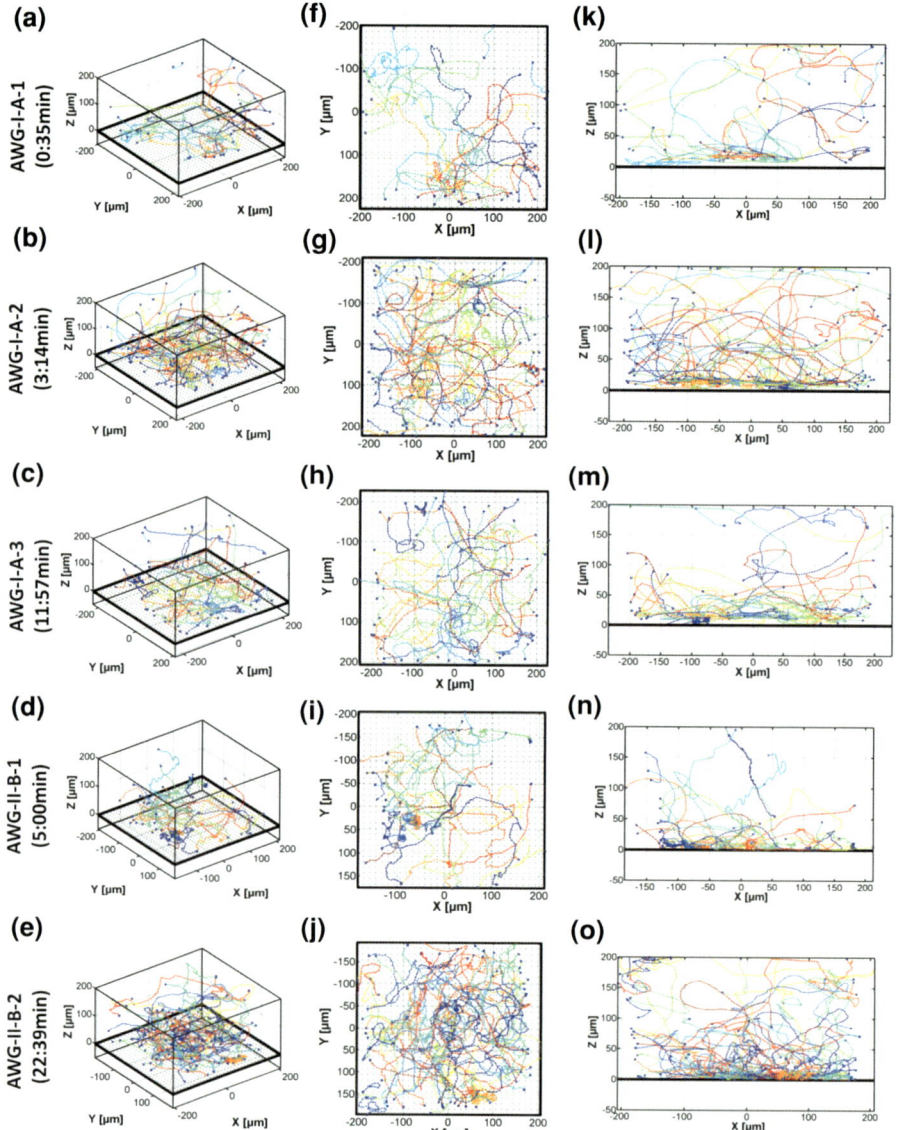

Fig. A.2 3D rendered plots of spores within a distance of 0–200 μm from the surface; **a–e** 3D view; **f–j** xy view; **k–o** xz view. To distinguish between trajectories they are marked in different colors

All trajectories can be divided into several phases (highlighted in each figure). The phases are named (I, II, III,…) according to their temporal occurrence. In Table A.3 these phases are analyzed in detail. In general the motility can be divided into a motion in solution, a motion in the boundary layer near the surface (20–50 μm) and a motion with surface contact.

Table A.2 Distribution of the determined motion patterns for AWG

Name	Time (min)	Gyration (*hit and run*)	Spinning	Undefined	Total
AWG-I-A-1	0:35	9 (3)	0	1	13
AWG-I-A-2	3:14	27 (8)	0	2	37
AWG-I-A-3	11:57	17 (5)	2	1	25
Sum		53 (16)	2	4	75
AWG-II-B-1	5:00	12 (6)	6	0	24
AWG-II-B-2	22:39	22 (9)	8	1	40
Sum		34 (15)	14	1	64

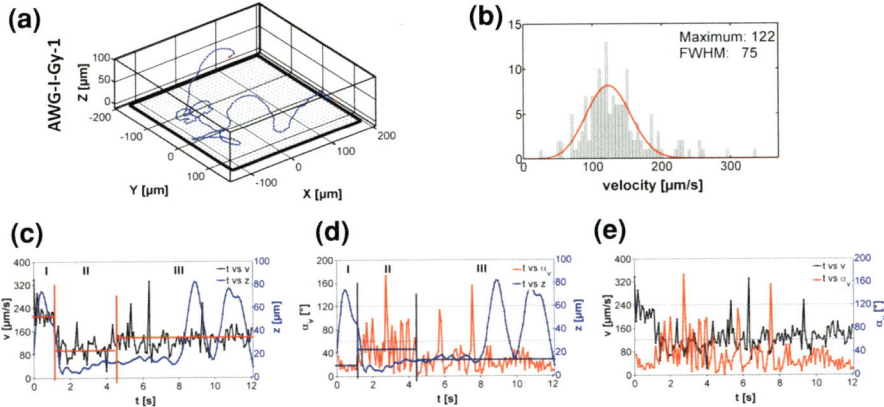

Fig. A.3 Example 1 (AWG-Gy-1) for the *gyration* pattern. **a** 3D rendered trajectory; **b** velocity histogram with fitted Maxwell–Boltzmann distribution; **c** velocity (*black, left side*) and distance to the surface (*blue, right side*) versus elapsed time; **d** α_v (*red, right side*) and distance to the surface (*blue, left*) versus elapsed time; **e** velocity (*black, right side*) and α_v (*red, left*) versus elapsed time

The values shown in Table A.3 and the corresponding Figs. A.3, A.4 and A.5 show that if a spore gets close to the surface the swimming behavior changes. In the approach or detachment phase (from the surface) the spore swims as fast as a spore assigned to the *orientation* pattern ($v_p > 100$ μm/s). The α_v distribution for this phase is also typical for an *orientation* pattern ($\bar{\alpha} \approx 30°$, see Sect. 5.1.3). If a spore gets close to the surface the α_v distribution changes significantly. The $\bar{\alpha}$ value is $60 \pm 30°$. The velocity also decreases when a spore gets close to the surface. In Figs. A.3, A.4 and A.5, panel (e) it can be observed that the decrease in speed and the increase in α_v occurs at the same time.

A.1.1.2   AWG Swimming Pattern: Hit and Run

The *hit and run* motion is a special case of the *gyration* pattern but it is listed as an individual pattern here, because it is important for the description of the

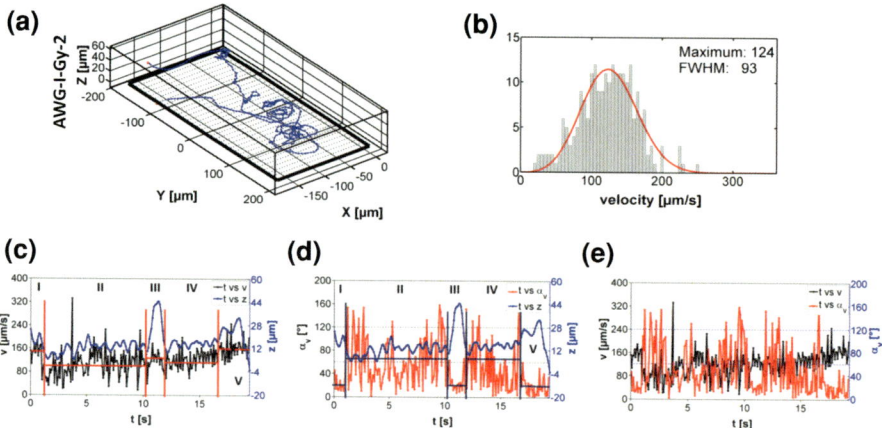

Fig. A.4 Example 2 (AWG-I-Gy-2) for the *gyration* pattern. **a** 3D rendered trajectory; **b** velocity histogram with fitted Maxwell–Boltzmann distribution; **c** velocity (black, left side) and distance to the surface (blue, right side) versus elapsed time; **d** α_v (*red, right side*) and distance to the surface (*blue, left*) versus elapsed time; e velocity (*black, right side*) and α_v (*red, left*) versus elapsed time

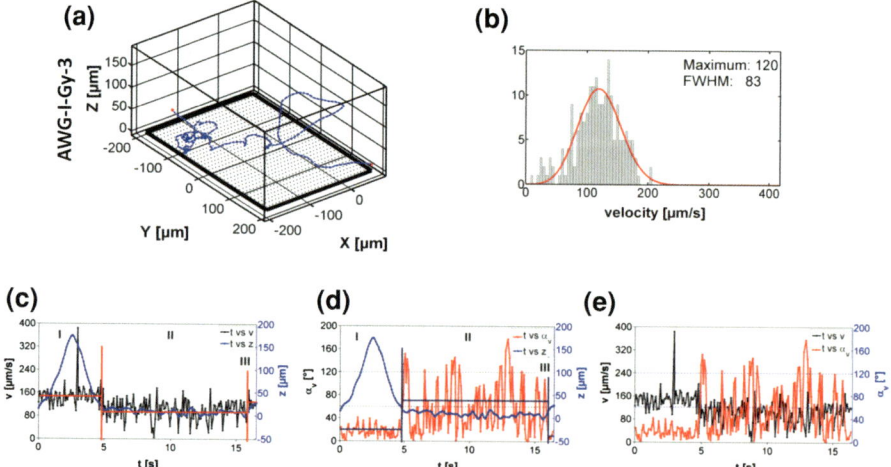

Fig. A.5 Example 3 (AWG-I-Gy-3) for the *gyration* pattern. **a** 3D rendered trajectory; **b** velocity histogram with fitted Maxwell–Boltzmann distribution; **c** velocity (*black, left side*) and distance to the surface (*blue, right side*) versus elapsed time; **d** α_v (*red, right side*) and distance to the surface (*blue, left*) versus elapsed time; **e** velocity (*black, right side*) and α_v (*red, left*) versus elapsed time

exploration behavior. Figs. A.6, A.7 and A.8 show three example trajectories for the *hit and run* movement. All shown spores move fast (>100 μm/s) towards and away from the surface. None of the shown spores stays close to the surface for a

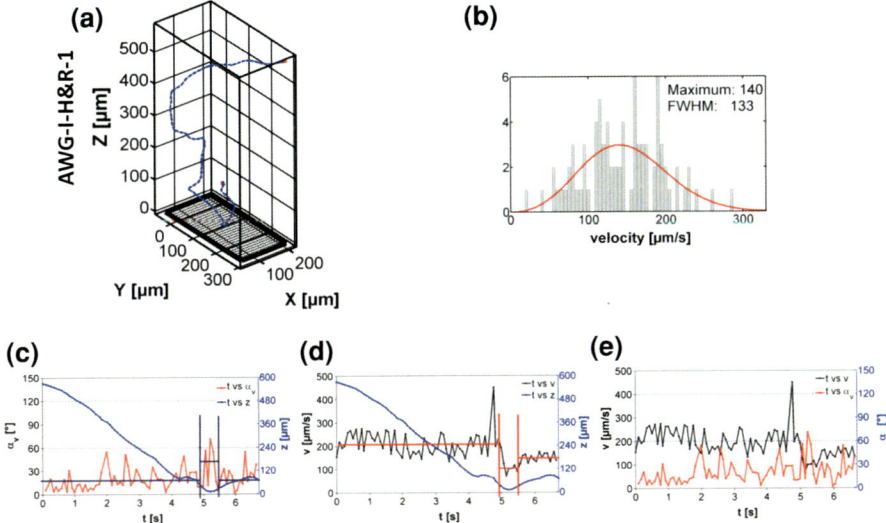

Fig. A.6 Example 1 (AWG-I-H&R-1) for a *hit and run* movement. **a** 3D rendered trajectory; **b** velocity histogram with fitted Maxwell–Boltzmann distribution; **c** velocity (*black, left side*) and distance to the surface (*blue, right side*) versus elapsed time; **d** α_v (*red, right side*) and distance to the surface (*blue, left*) versus elapsed time; **e** velocity (*black, right side*) and α_v (*red, left*) versus elapsed time

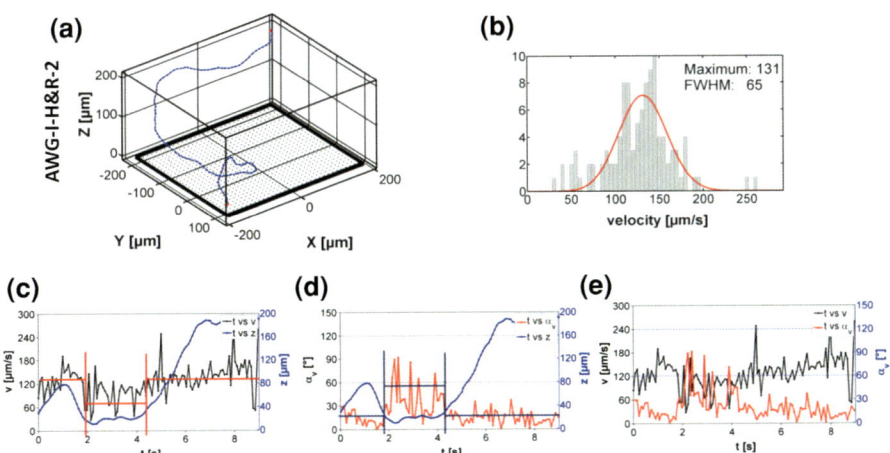

Fig. A.7 Example 2 (AWG-I-H&R-2) for a *hit and run* movement. **a** 3D rendered trajectory; **b** velocity histogram with fitted Maxwell–Boltzmann distribution; **c** velocity (*black, left side*) and distance to the surface (*blue, right side*) versus elapsed time; **d** α_v (*red, right side*) and distance to the surface (*blue, left*) versus elapsed time; **e** velocity (*black, right side*) and α_v (*red, left*) versus elapsed time

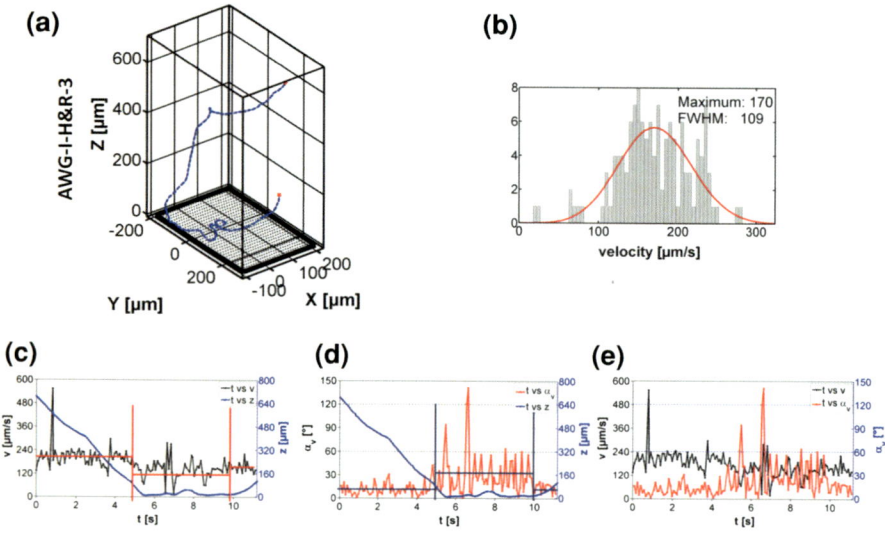

Fig. A.8 Example 3 (AWG-I-H&R-3) for a *hit and run* movement. **a** 3D rendered trajectory; **b** velocity histogram with fitted Maxwell–Boltzmann distribution; **c** velocity (*black, left side*) and distance to the surface (*blue, right side*) versus elapsed time; **d** α_v (*red, right side*) and distance to the surface (*blue, left*) versus elapsed time; **e** velocity (*black, right side*) and α_v (*red, left*) versus elapsed time

Fig. A.9 Detailed settlement attempt of the spore shown in Fig. A.10. S1–S3 are the marked *spinning* phases

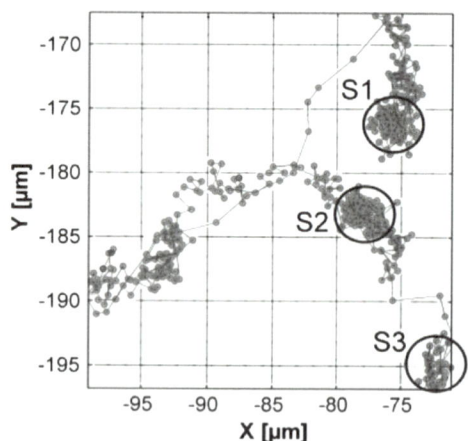

long time. In Figs. A.6, A.7 and A.8 plot (b) shows a velocity histogram for each spore. The velocity distribution is not significantly different from the distribution for a movement in solution (see Sect. 5.1.2).

The angle distribution (α_v) in Figs. A.6, A.7 and A.8 panels (d, e) is typical for a spore assigned to the *orientation* pattern. When the spores get closer to the surface,

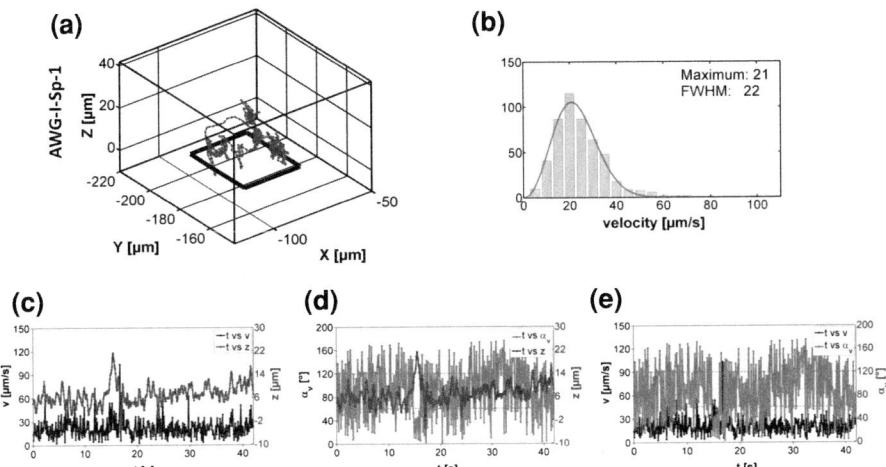

Fig. A.10 Example 1 (AWG-I-Sp-1) for a settlement attempt. **a** 3D rendered trajectory; **b** velocity histogram with fitted Maxwell–Boltzmann distribution; **c** velocity (*black, left side*) and distance to the surface (*gray, right side*) versus elapsed time; **d** α_v (*gray, right side*) and distance to the surface (*black, left*) versus elapsed time; **e** velocity (*black, right side*) and α_v (*gray, left*) versus elapsed time

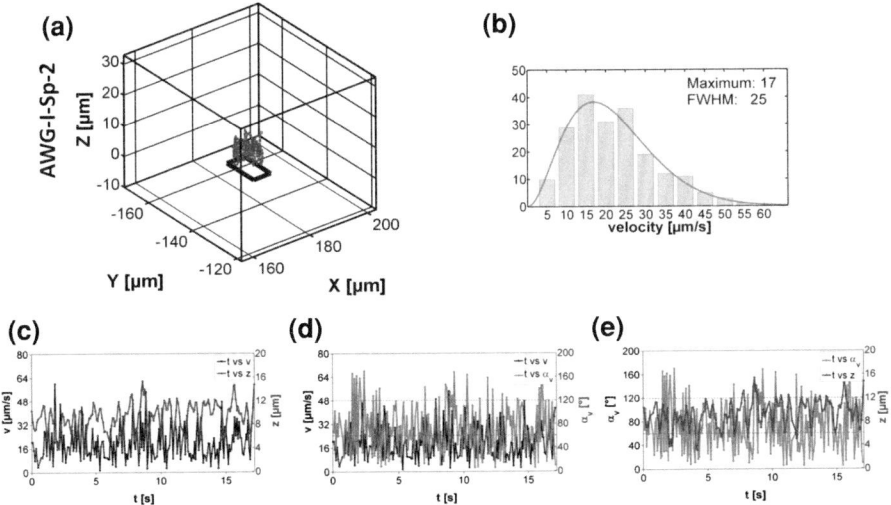

Fig. A.11 Example 2 (AWG-II-Sp-2) for a settlement attempt. **a** 3D rendered trajectory; **b** velocity histogram with fitted Maxwell–Boltzmann distribution; **c** velocity (*black, left side*) and distance to the surface (*gray, right side*) versus elapsed time; **d** α_v (*gray, right side*) and distance to the surface (*black, left*) versus elapsed time; **e** velocity (*black, right side*) and α_v (*gray, left*) versus elapsed time

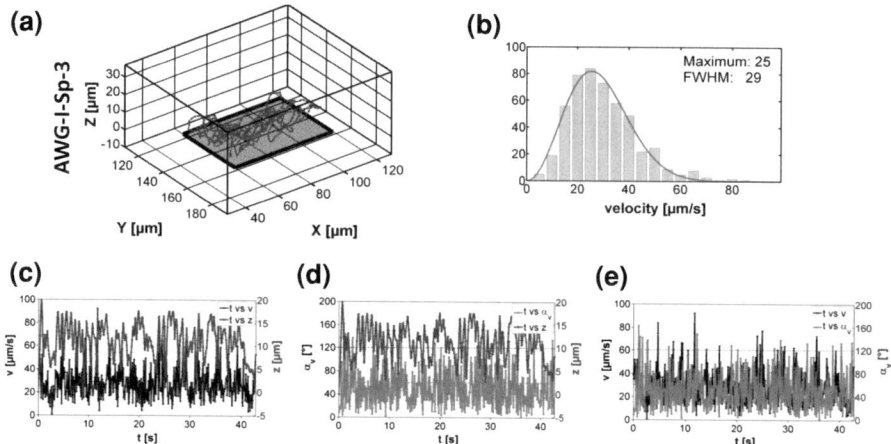

Fig. A.12 Example 3 (AWG-II-Sp-3) for a settlement attempt. **a** 3D rendered trajectory; **b** velocity histogram with fitted Maxwell–Boltzmann distribution; **c** velocity (*black, left side*) and distance to the surface (*gray, right side*) versus elapsed time; **d** α_v (*gray, right side*) and distance to the surface (*black, left*) versus elapsed time; **e** velocity (*black, right side*) and α_v (*gray, left*) versus elapsed time

Table A.3 Z-position and time if the spore change it motility

AWG-I-Gy-1	Phase I	Phase II	Phase III		
t (start) (s)	0.0	1.0	4.6		
t (end) (s)	1.0	4.6	12.1		
z (start) (µm)	20	46	13		
z (end) (µm)	46	13	9		
v_m (µm/s)	210	85	130		
$\bar{\alpha}$ (°)	15	45	30		
AWG-I-Gy-2	Phase I	Phase II	Phase III	Phase IV	Phase V
t (start) (s)	0.0	1.0	10.3	11.9	19.7
t (end) (s)	1.0	10.3	11.9	16.7	19.4
z (start) (µm)	25	15	10	12	20
z (end) (µm)	15	10	12	20	0
v_m (µm/s)	155	100	120	110	155
$\bar{\alpha}$ (°)	15	70	15	70	20
AWG-I-Gy-3	Phase I	Phase II	Phase III		
t (start)1 (s)	0	14.3	15.8		
t (end) (s)	14.3	15.8	16.5		
z (start) (µm)	5	17	2		
z (end) (µm)	17	2	28		
v_m(µm/s)	158	90	158		
$\bar{\alpha}$ (°)	20	70	20		

the α_v distribution on average has a higher value and shows more fluctuations. The value of α_v fluctuates around $60 \pm 30°$, sometimes even higher. In Fig. A.7 in panels (d, e) the increase in α_v is clearly visible for the time span (1.8–4.5 s) while the spore is close to the surface. The change in α_v occurs when the spore swims closer to the surface than 40 µm. When it leaves the surface the α_v distribution is typical for a "free" swimming spore at a distance of ≈ 30 µm from the surface. In Fig. A.6 this effect is not pronounced that strongly because the spore is close to the surface only for an extremely short time. But for this time span (5.0–5.5 s) the angular distribution is larger as for the spore swimming in a larger distance from the surface. The change in the α_v distribution occurs when the spore is closer than 22 µm to the surface. The distribution is—again—typical for a "free" swimming spore when the spore swims further than 30 µm away from the surface.

For the spore represented by the trajectory shown in Fig. A.8 the increase in α_v is also clearly visible. The spore stays in close contact to the surface for quite a long time (4.8–10.0 s) and swims parallel to it before it leaves the FoV in a distance of 150 µm away from the surface. The first change of α_v is detected 100 µm from the surface and the α_v distribution is "normal" for a swimming spore in solution when the spore leaves the surface in a distance of 17 µm from the surface.

In Figs. A.6, A.7 and A.8 panels (c, e) the velocity (black, left side) is plotted versus the elapsed time. Additionally in panel (d) the dependence of the distance towards the surface (blue, right side) and in panel (e) α_v (red, right side) is included. These plots also show that the velocity of the spore movement is influenced by the presence of the surface. The velocity decreases when the spore gets close to the surface. The decrease in speed and the increase in α_v occur at the same time [see Figs. A.6, A.7 and A.8, panel (e)].

A.1.1.3 AWG Swimming Pattern: Spinning

Before a spore settles it spins on the surface for various time spans. The connection between settlement and *spinning* has already been explained in Sect. 5.2.2 and Sect. 5.2.3 in the context of the different identified exploration patterns. The expected settlement on glass within a holographic experiment is two spores per 40 min in the FoV. This settlement rate is confirmed in the hologram analysis. The *spinning* on AWG is similar to the *spinning* on FOTS, but on AWG it always occurs subsequently to the *gyration* pattern. The pattern is not as distinctive as on FOTS because no spore has yet selected the surface suitable to settle within the analyzed time. In Heydt et al. [1] a more pronounced *spinning* event on AWG is already shown which, back then, was entitled "search pattern" (instead of *spinning*). Nevertheless the first short spinning phases are observed.

In Fig. A.9 the top view of the spores AWG-I-Sp-1 is shown. Three individual *spinning* phases are observed. Between these phases the spores do not leave the surface and stays close to it [see also Fig. A.10, panels (a, c, d)]. After several *spinning* attempts (S1–S3) the spore leaves the FoV. As defined in Sect. 5.2.4.5 during the *spinning* phase the spore is fixed a position on the surface and rotates around this position. During this motion the v_m and α_v values do not have any physiological meaning. Therefore the spinning motion is characterized by the spinning radius (ra) and the angle β (see Fig. 4.18). Since, the *spinning* phase presented here only last for less than a second and subsequently the spores moves parallel to the surface the trajectory is characterized in the same way as for the *gyration* and *hit and run* pattern. The values of v_m and α_v shown in the following Figures are only used to discriminate the *spinning* phase from the motility along the surface and in solution without introducing any physiological to the value. The *spinning* phase itself is analyzed in detail in Sect. A.3.1.2.

In total 14 individual trajectories are assigned to the *spinning* pattern (see Table A.2). The *spinning* pattern is a rare event on AWG in the small FoV. All occurring events are analyzed. The spores shown in Figs. A.10, A.11 and A.12 swim with a constant but slow average velocity (15–25 µm/s). The spinning phases at this point in time only last for a short period. Subsequently the spore travels a small distance along the surface to start the next *spinning* phase. During the complete trajectory the α_v distribution fluctuates around $77 \pm 42°$ so that the spinning phases and the movement phases are not easy to separate. Also the Z-position changes only within the length scale of the spore body (± 5 µm) which is typical for the *spinning* pattern.

A.1.2 *Exploration on AWG: General Behavior*

On AWG the spore motility in the vicinity of the surface can be described by the swimming pattern: *gyration* (occurs in majority), *hit and run* and *spinning*. In Table A.2 the occurrence of the patterns is summarized. To obtain a more general understanding of the exploration behavior velocity histograms are analyzed. One aspect of the following discussion is to verify the observations of the last section— shown there for exemplary traces—for all recorded spores. Therefore, most of the important observations of the last section are summarized.

- Spores which are assigned to the *gyration* pattern swim fast ($v_p > 100$ µm/s) if they are far from the surface. For this part of the trajectory the velocity and α_v distribution is similar to the *orientation* pattern.
- Spores are slower if they are close to the surface. Based on the analysis of the last section, the decrease of the velocity should be observable at least 50 µm from the surface.

To simplify the complex analysis, histograms are first calculated only for three different sections of the observation volume. Afterwards the whole observation volume is analyzed in greater detail. The sections are defined according to the distance from the surface and named in the following manner: *bulk* 200–800 µm, *near* the surface 50–200 µm and *close* the surface 0–50 µm. The definitions of the sections are based on the observations of the last chapter. The *bulk* behavior has been discussed in Sect. 5.1.

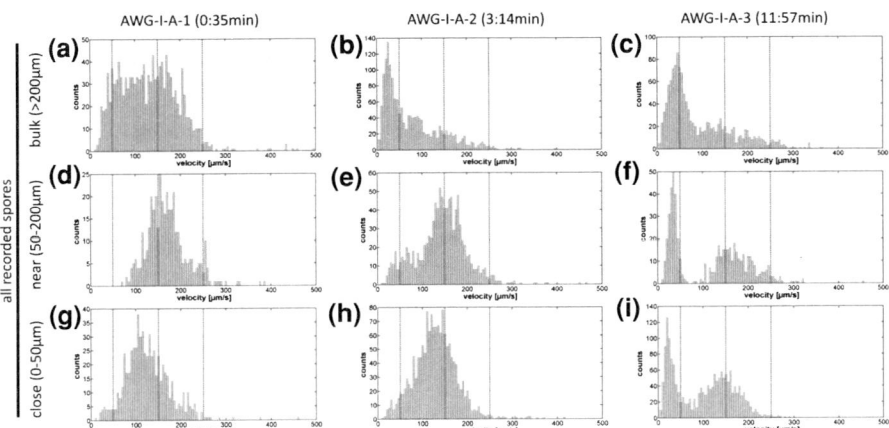

Fig. A.13 Velocity histogram for AWG-I-A-* shown for three distances (**a–c** 200–800 μm, **d–f** 50–200 μm and **g–i** 0–50 μm) away from the surface. (**a, d, g**) is analyzed after 0:35 min, (**b, e, h**) after 3:14 min and (**c, f, i**) after 11:57 min. The data shown in (**a–c**) is already discussed in Sect. 5.1.1 under the name Bulk-I-A-*. The *vertical lines* indicate a velocities of 50, 150, 250 μm/s

The histograms shown in Fig. A.13 include all recorded spores. Figure A.14 shows the histogram distributions for spores assigned to the *gyration* [panels (d–i)], *spinning* [panels (j–l)], and *orientation* [panels (a–c)] patterns separately. Both figures are arranged in the same order: *bulk* [200–800 μm, panels (a–c)], *near* the surface [50–200 μm, panels (d–f) and *close* the surface (0–50 μm), panels (g–i)]. The spinning pattern only occurs *close* to the surface. The histograms in Figs. A.13 and A.14 are furthermore shown for different point in times. The histograms in panels (a, d, g, j) are recoded after 0:34.5 min, panels (b, e, h, k) after 3:13.7 min and panels (c, d, i, l) after 11:57.0 min. The swimming performance in the *bulk* [panels (a–c)] is already explained in Sect. 5.1.1 under the name Bulk-I-A-*.

The following discussion refers to the Figs. A.13 and A.14. Shortly after the injection [0:34.5 min, Fig. A.13 panels (a, d, g)] only spores assigned to the *orientation* and *gyration* pattern are *close* [0–50 μm, panel (g)] and *near* [50–200 μm, panel (d)] the surface, whereas in the *bulk* [>200, panel (a)] the *wobbling* and *orientation* pattern dominates. This statement is confirmed when the distribution for all spores [Fig. A.13, panels (a, d, g)] is compared to the histograms for the spores assigned to the *gyration* pattern [Fig. A.14, panels (a, d, g)]. In the histogram *close* to the surface slow velocity values are predominant. Comparing the histogram for all spores [Fig. A.13, panel (g)] to the spores assigned to the *gyration* pattern [Fig. A.14, panel (g)] no significant differences are detected. Therefore all slow velocity values in the histogram are due to spore assigned to the *gyration* pattern and have to be due to interaction of spores with the surface (see exemplary traces in Sects. A.1.1.1 and A.1.1.2).

With elapsing time *wobbling* spores occur in nearly the whole volume. The amount of slow spores *near* the surface [panels (e, f)] increases with elapsing time. In the intermediate time [panels (b, e, h)] a few "slow" spores are detected *near*

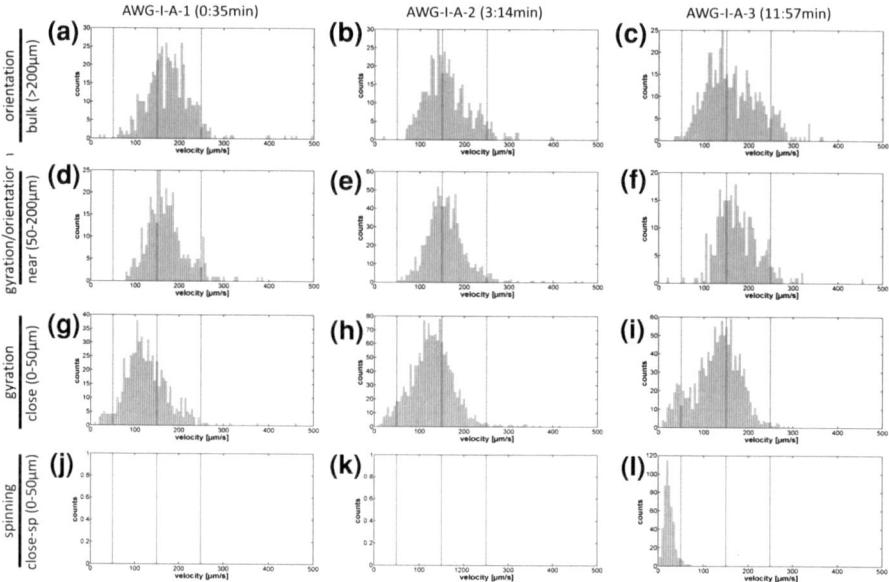

Fig. A.14 Velocity histogram on AWG-I-A-* for the spores assigned to the fast spore fraction in the bulk **a–c**, to the *gyration* **d–i** and *spinning* **j–l** pattern. The histograms are shown for three distances **a–c** 200–800 μm, **d–f** 50–200 μm and **g–l** 0–50 μm away from the surface. **a, d, g, j** is analyzed after 0:35 min, **b, e, h, k** after 3:14 min and **c, f, i, l** after 11:57 min. The data shown in **a–c** is already discussed in Sect. 5.1.1 under the name Bulk-I-A-*. The *vertical lines* indicate a velocities of 50, 150, 250 μm/s

the surface [panel (e)], but no "slow" spores are detected *close* to the surface [panel (h)], even though some slow velocity values are present in the histogram. However, these values are not caused by *wobbling* spores because there is no significant difference detected when the velocity histograms shown in Fig. A.13, panel (h), all recorded spores and Fig. A.14 panel (h), only *gyration* are compared. This leads to the conclusion that the slow velocity values in the histograms are due to spores assigned to the *gyration* pattern and thus interaction with the surface. The situation *near* (50–200 μm) the surface [panel (e)] is different at this intermediate point in time. When the histograms for all spores [Fig. A.13, panel (e)] and the spores only assigned to the *gyration* and *orientation* pattern [Fig. A.14, panel (e)] are compared the small peak in Fig. A.13, panel (e) can be assigned to spores belonging to the *wobbling* pattern.

At the last analyzed point in time [11:57 min, panels (c, f, i)], the "slow" spores are present in a majority in nearly the complete observation volume [panels (c, f)]. But still, *close* to the surface the *wobbling* pattern is not observed. The situation at this point in time is more complex because the second surface exploration pattern (*spinning*) also occurs [see Fig. A.14, panel (l)]. As on AWG-II-B-* the first spinning event is observed after 3:33 min (result not shown as an individual figure) it cannot be generalized that the *spinning* occurs so late on AWG-I-A-*. If the histogram for the *spinning* motion [Fig. A.14, panel (l)] and the histogram for the spores assigned to the *gyration* pattern [Fig. A.14, panel (i)] are combined together this "sum" represents the histogram for all spores [Fig. A.13, panel (i)]. The conclusion is that spores belonging to the *wobbling* pattern are *close* to the surface and therefore these spores do not actively explore the surface. All exploration events up to this time can be assigned to spores swimming in the *gyration* or *spinning* pattern.

If the histograms for spores assigned to the *gyration* pattern (Fig. A.14) are compared after 0:34.5 min [panel (g)], after 3:13.7 min [panel (h)] and after 11:57 min [panel (i)] the slow velocity values increase with elapsing time. This leads to the conclusion that the number of surface interactions also increases. The most probable velocity (v_p) for the spore assigned to the *gyration* pattern [Fig. A.14] is in a good agreement for the spore in the *bulk* [panels (a, b, c)] and *near* [panels (d, e, f)] to the surface. But—particularly in the beginning of the experiment [panel (g)]—v_p *close* to the surface [panels (g, h, i)] is significantly slower than *near* the surface [panels (d, e, f)]. This means that the spores *close* to the surface are slower than in the *bulk*. To study this effect in detail the sectioning of the observation volume in only three sections is too crude. A detailed analysis is provided after the general spore distribution in the observation volume is discussed.

In Fig. A.15 the general spore distribution is shown for the complete observation volume. The distribution is shown as histograms for all spores [panels (a–c)] and for spores only assigned to the *gyration* and *orientation* pattern [panels (d–f)]. The distribution is also shown for the different analyzed points in time. The distributions are all fairly similar to each other and the same trend can be found in each distribution. Spores accumulate *close* to the surface. This enrichment is detected starting at a distance of 160 μm from the surface. Even if the first *spinning* pattern is not observed before the last analyzed experiment (AWG-I-A-3) the surface accumulation is already developed. The distribution in the first

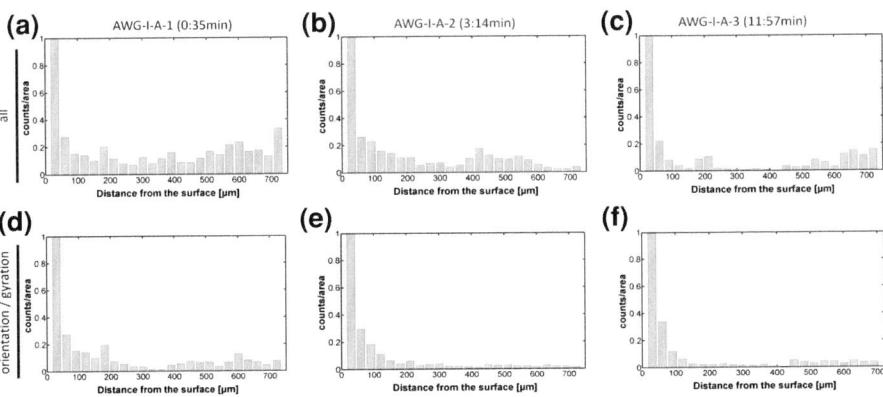

Fig. A.15 General spore distribution on AWG-I-A-* in the complete observation volume. **a–c** all recorded spores, **d–f** spores only assigned to the *gyration* and *orientation* pattern. **a, d** is recorded after 0:35 min, **b, e** after 3:14 min and **c, f** after 11:57 min

experiment [AWG-I-1, panels (a, d)] is identically for the first six columns (180 μm) for all analyzed spores [panel (a)] and the spores assigned to the *orientation* and *gyration* pattern [panel (d)]. This means that the *wobbling* pattern is not determined *near* the surface. At the second point in time [AWG-I-2, panels (b, e)] only the first 5 bins of the histogram are the same. The slow spore fraction gets closer to the surface but is not responsible for the surface enrichment. In the last experiment the situation is more complex because of the occurrence of the *spinning* pattern. The spores assigned to the *spinning* pattern just influence the height of the first bin [panel (c)]. The bins 2–4 are only populated by spores assigned to the *gyration* pattern [see comparison panels (c, f)]. The conclusion of this discussion is that the spores assigned to the *gyration* pattern are responsible for the accumulation close to the surface.

Based on the results for the velocity histograms (Figs. A.13, A.14) and the spore distribution (Fig. A.15) it can be conducted that the spores assigned to the *wobbling* pattern do not explore the surface and the area close to the surface. A spore assigned to the *spinning* pattern actively explores the surface but does not provide information about how the spore swam to this specific surface position. Therefore, to study the exploration behavior of spores swimming towards and away from the surface only spores assigned to the *gyration* and *orientation* patterns can provide information. However, for completeness and to verify this statement the distribution for all spores is also provided. The spore distribution in the observation volume for the spores assigned to the *gyration* pattern [Fig. A.15, panels (d, e, f)] does not change significantly with elapsing time. Therefore all spores from all experiments (AWG-I-A-*) are analyzed together.

In Figs. A.16, A.17, A.18 and A.19 the surface exploration behavior for AWG is analyzed. From now on the observation volume is analyzed by dividing the volume in equidistant sections parallel to the surface. Each section has a height of 30 μm. For each section the data is plotted as a histogram. The histograms of each section are stacked together and combined in one graph. In Fig. A.16, panels (a, d, g, j) a 3D view is shown. The bars in the graph are colored according to their height, the higher the bar the darker the color. Fig. A.16, panels (b, e, h, k) gives a 2D top view of the perspective 3D plots. Darker squares represent higher bars. Fig. A.16, panels (c, f, i, l) shows the mean value for each section of the variable shown in the panel before.

Figure A.16 shows the angle distribution (α_v) for all spores [panels (a–c)], for the spores assigned to the *gyration* and *orientation* pattern [panels (d–f)]. Also in Fig. A.16 the α_z distribution is shown. The angle α_z is the angle of the spore velocity vector with respect to the surface normal and is illustrated in Fig. 4.15, panel (b). If α_z is smaller than 90° the spore swims away from the surface and if α_z is bigger than 90° the spore swims towards the surface. In Fig. A.16 the α_z distribution is shown for all analyzed spores [panels (g–i)] and for the spores assigned to the *gyration* pattern only [panels (j–l)]. Fig. A.16, panels (a–f) shows that close to the surface the α_v distribution is extremely variable and many high values occur. This means that spores change their swimming direction abruptly and frequently. In solution fewer changes in the swimming direction occur and therefore the value

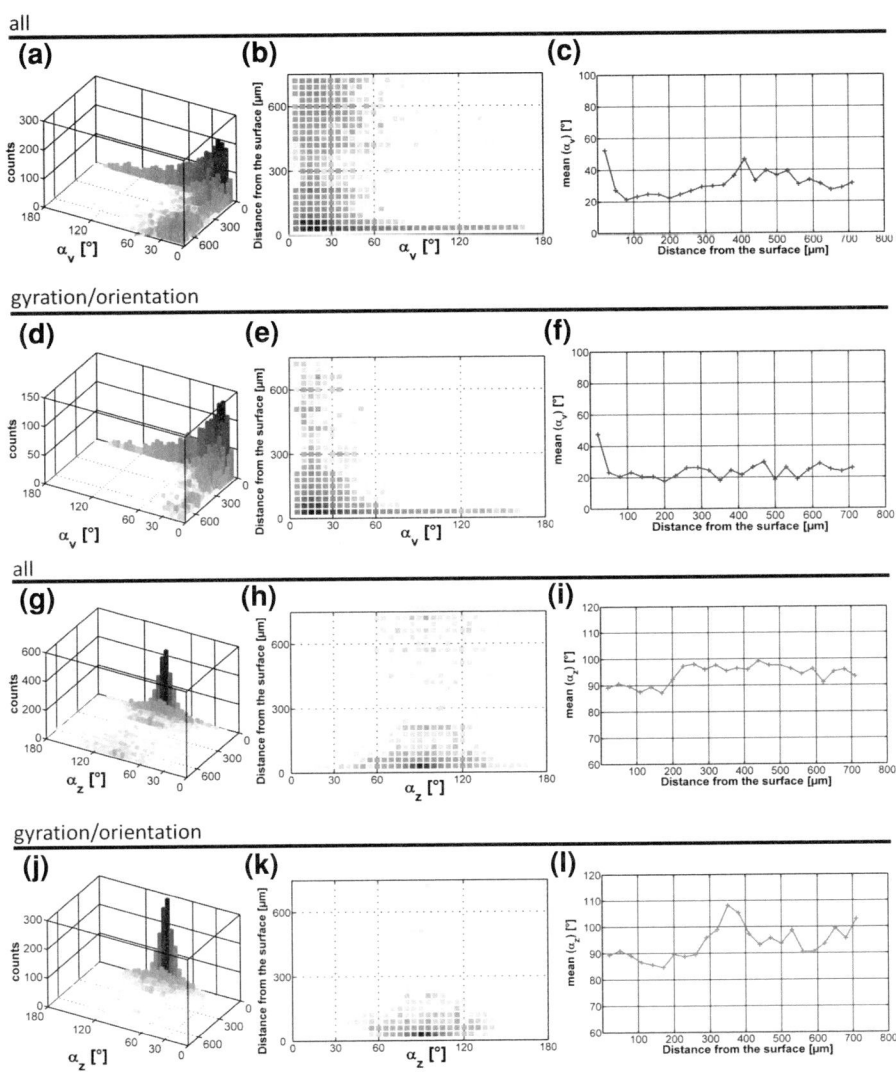

Fig. A.16 α_v and α_z distribution on AWG-I-A-*. **a, d, g, j** 3D histogram; **b, e, h, k** 2D top view of the perspective 3D histogram; **c, f, i, l** mean value of value shown in the two panels before; **a–c** α_v for all recorded spores; **d–f** α_v for the spores only assigned to the *gyration* pattern, **g–i** α_z for all recorded spores and **j–l** α_z for the spores assigned to the *gyration* pattern. The higher the numbers of counts the darker the color of the bar

of the α_v distribution is smaller (see Sect. 5.1). In Fig. A.16, panels (a–f) it is possible to analyze at which distance from the surface the α_v distribution observed in solution changes into the α_v distribution found close to the surface. The effect is more pronounced for the spores assigned to the *gyration* pattern [panels (d–e)] than for all recorded spores. As described earlier (see Sect. 5.1.2.2) the spores

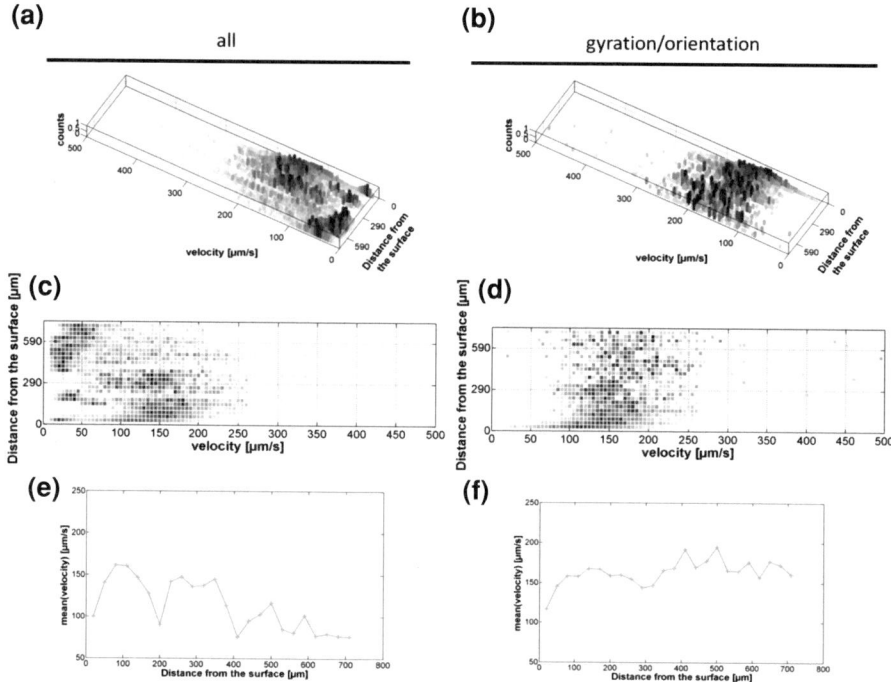

Fig. A.17 Velocity distribution on AWG-I-A-*. **a, c, e** for all recorded spores; **b, d, f** for the spores assigned to the *gyration* pattern; **a, b** 3D histogram; **c, d** 2D top view of the perspective 3D histogram; **e, f** mean value of the corresponding panel

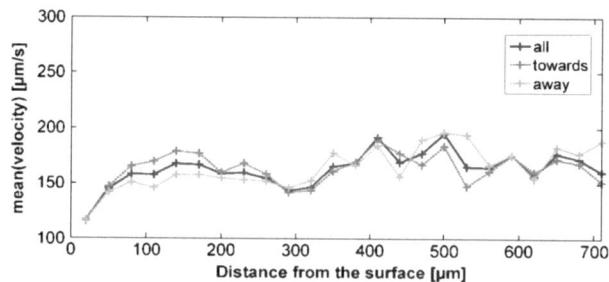

Fig. A.18 Detailed analysis of $\bar{v}$ for AWG-I-A-*. The spores shown are assigned to the *gyration* and *orientation* pattern. These spores are separated in the spores swimming towards the surface (*gray*), all spores (*black*) and spores swimming away from the surface (*light gray*)

assigned to the wobbling pattern swim more fidgety than the spore fraction assigned to the *orientation* pattern and therefore the α_v distribution for all spores [see panels (a–c)] is more changeful in solution.

At a distance of 200 μm panel the surface the distribution which is observed for the solution starts to change towards the distribution *close* to the surface [see Fig. A.16,

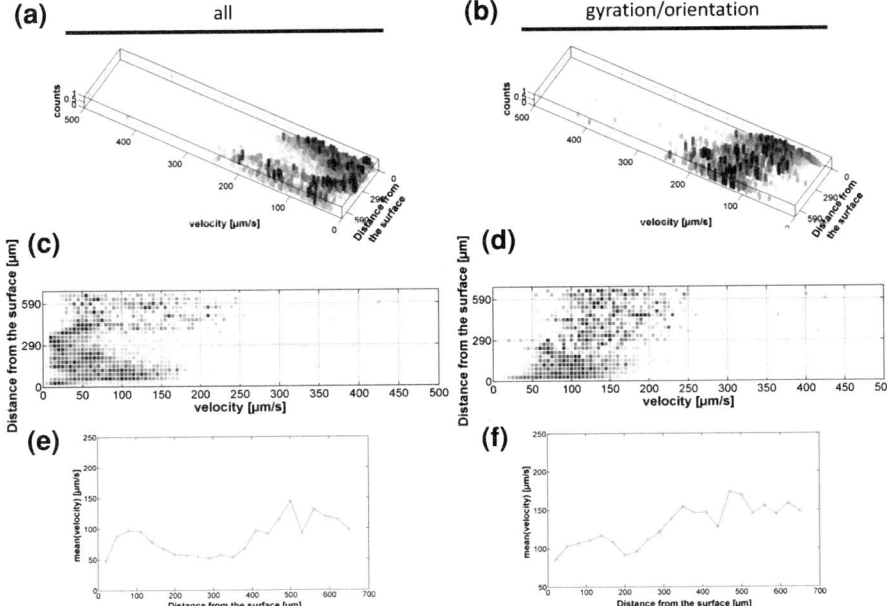

Fig. A.19 velocity distribution AWG-II-B-*. **a, c, e** for all recorded spores; **b, d, f** for the spores assigned to the *gyration* pattern; **a, b** 3D view; **c, d** xy view; **e, f** mean value of the corresponding panel before

panel (e)]. For the mean value of α_v the change is also observed but only at a distance of 90 µm away from the surface. The change in α_v is also observed for all recorded spores [see panels (b, c)]. For the mean distribution [panel (c)] the change occurs at the same position as for the spore assigned to the *gyration* pattern [panel (f)].

In Fig. A.16, panels (g–l) the α_z distribution is shown. In solution the distribution is broad and no maximum is detected. That means there is no large preference to swim in a certain spatial direction. Close to the surface the spores swim with a strong preference for 90° which means that the spores swim parallel the surface. The $\bar{\alpha}_z$ distribution for all spores [panel (i)] shows that from a distance of 200 to 720 µm from the surface the spores prefer (slightly) to swim towards the surface, because the mean value of distribution is 98°. For the spores only assigned to the *gyration* pattern this trend is also observed. The peak between 300 and 400 µm is an outlier due to the statistics in this z-regime. The low statistic for this Z-regime is clearly visible in Fig. A.16, panels (j, k). Within the first 200 µm from the surface [panels (i, l)] the same number of spores swims away and towards the surface. This equilibrium is due to the exploration behavior of the spores. In the *gyration* pattern the spores swim up and down and search for a place to settle. The spores which are "lost" for the equilibrium because of settlement or spinning are replaced by spores coming from solution.

In Fig. A.17 the change in the velocity distribution is investigated for the complete volume. The observation volume is sectioned as for Figs. A.16, A.17, panels (a, c, e) shows the distribution for all spores and Fig. A.17, panels (b, d, f) for the spores assigned to the *gyration* and *orientation* pattern. The spores assigned to the *gyration* and *orientation* pattern swim slower when they are close to the surface. This observation is already stated in the discussion of Fig. A.14. In Fig. A.17 the change of the velocity in solution and close the surface can be investigated. The slowdown can be detected at a distance of 120 μm from the surface as well as in the 2D top view of the histogram as for the mean value [see panels (d, f)]. For all recorded spores the slowdown is not so easy to determine, because of the occurrence of the *wobbling* pattern [see panels (a, c, e)].

In Fig. A.18 the velocity dependency for spores swimming towards and away from the surface is studied in detail for the spores assigned to the *gyration* and *orientation* pattern. The distribution is divided in three sections: all spores (black), spores which swim towards the surface (gray) and spores which swim away from the surface (light gray). The discrimination whether a spore swims towards or away from the surface is based on the α_z value. All α_z-values bigger than 90° mean that a spore is swimming towards the surface and every value of α_z smaller than 90° means that a spore is swimming away from the surface. In the first 200 μm from the surface the spores swim away from the surface slower than towards the surface. In bulk the spores swimming away or towards the surface are equally fast.

In Figs. A.19 and A.20 the same analysis is provided for the spores analyzed at collection trip B (AWG-II-B-*). The data for this experiment is analyzed after 3:33 min and 22:38 min. Both examinations are combined as it is done for AWG-I-A-*. Also the observation volume is divided into the same sections as in AWG-I-A-*.

In Fig. A.20, panels (a–c) α_v for all recorded spores are shown, panels (d–f) shows the α_v distribution for the spores assigned to the *gyration* and *orientation* pattern, panels (g–i) shows the α_z distribution for all spores and panels (j–l) shows the α_z distribution of all spores assigned to the *gyration* and *orientation* pattern.

The α_v distribution starts to change from the solution distribution to the surface distribution at a distance of 200 μm from the surface [panel (e)]. This distance is also observed for $\bar{\alpha}_v$ shown in Fig. A.20, panel (f). The observation for α_z of all recorded spores [Fig. A.16, panel (i)] in experiment AWG-I-A-* is that for the first 200 μm from the surface the same amount of spores swim away and towards the surface. With larger distances the spores have a preference to swim towards the surface. This observation is also—less pronounced—found for AWG-II-B-* [Fig. A.20, panel (i)]. The reason why the effect smaller is might be due to the fact that the motion of *Ulva* spores is analyzed about 10 min later as for AWG-II-B-* than for AWG-I-A-*.

Even if the general performance of the spores used in AWG-I-A-* and AWG-II-B-* is different regarding the distribution of "slow" and "fast" spores and v_p of the fast spore fraction (see Figs. 5.3, A.17, A.19), the distance from the surface where the solution behavior changes to the surface exploration behavior is the same. The surface interaction distance is determined to be observable for a distance of ≈ 200 μm from the surface.

all

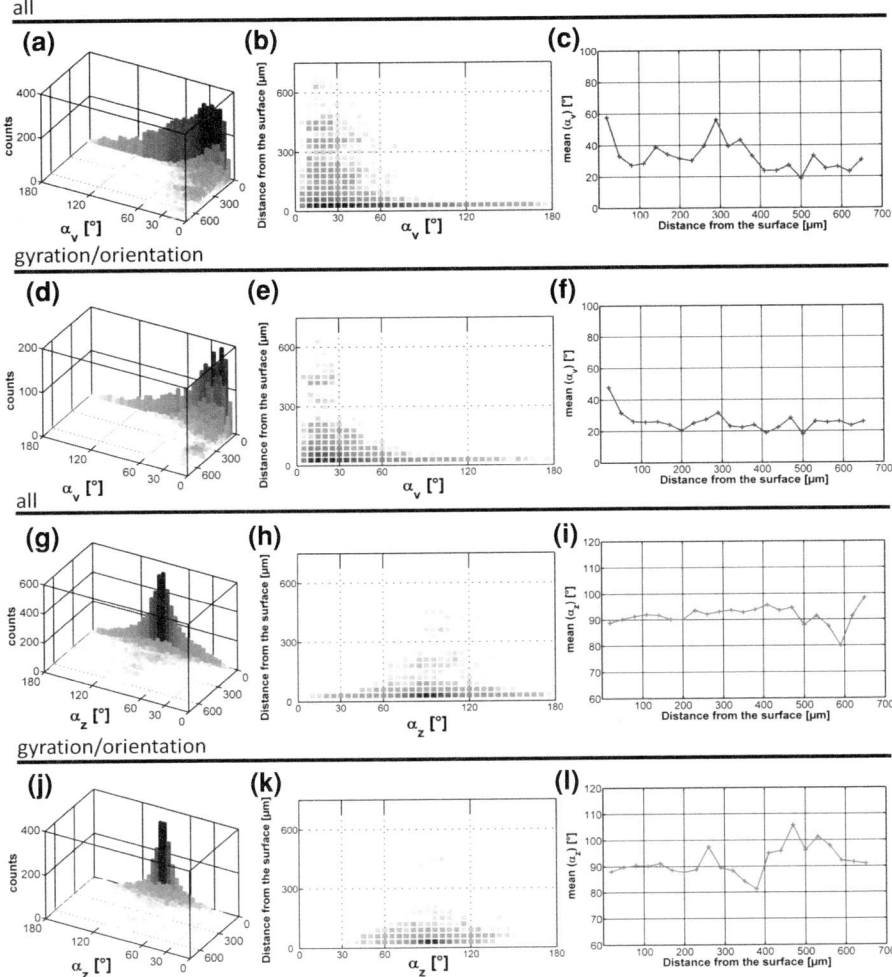

Fig. A.20 α_v and α_z distribution on AWG-II-B-*. **a, d, g, j** 3D histogram; **b, e, h, k** 2D top view of the 3D histogram; **c, f, i, l** mean value of value shown in the two panels before; **a–c** α_v for all recorded spores; **d–f** α_v for the spores only assigned to the *gyration* and *orientation* pattern; **g–i** α_z for all recorded spores and **j–l** α_z for the spores assigned to the *gyration* and *orientation* pattern

A.2 Exploration on PEG Coating

After the study of the exploration behavior on AWG in the previous section the exploration patterns of spores exploring PEG are analyzed. For spore settlement PEG is an unattractive (small settlement amount) surface and no settlement was observed in the FoV on this coating (see Sect. 5.2.2). This chapter is arranged in the same order as the previous. At first, the individual motility is discussed

Table A.4 Statistic for the exploration behavior on PEG

Name	Distance				Elapsed time (min:s)	Duration (s)
	0–50 μm		50–200 μm			
	Number of traces	Number of data points	Number of traces	Number of data points		
PEG-A-1	16	449	20	325	1:26	43.2
PEG-A-2	18	495	22	419	2:09	41.8
PEG-A-3	12	403	15	330	2:51	42.3
PEG-A-4	24	1,067	30	1,043	6:24	42.2
Sum	70	2,414	87	2,117		169.5

followed by analysis of the general behavior is analyzed. The following analysis is based on 98 individual traces and 4,544 data points. The available statistic is summarized in Table A.4. The experiment is named according to the previously used and explained systematic. The first recording is not before 1:26 min. Any time prior to this was not possible to analyze because of too much convection in the observation chamber. The convection is due to the heat of the CCD chip of the camera and has already been discussed in detail in Sect. 4.1.5.

In Figs. A.21 and A.22 the motility data of all recorded spores is shown as a 3D-rendered plot. Figure A.22 illustrate a magnification of motility *near* the surface (0–200 μm) of the complete observation volume which is shown in Fig. A.21. The typically erratic, random spore motion is observed for the exploration on PEG. No swarm behavior or convection is detected in the data. The spores accumulate within the first 200 μm from the surface. With elapsing time the amount of spores near the surface increases slightly (see Fig. A.21). The xz view in Fig. A.22, panels (i–l) shows that most spores—besides from one shown in panel (l)—do not swim down towards the surface. Most spores stop the approach to the surface in a distance of 5 μm away from the surface and turn around or swim along the surface only. Just one spore of the 98 recorded trajectories swims closer to the surface [panel (l) dark blue trace] and stays for a longer time at the interface. The motility characteristic of this trace is unusual with respect to the other recorded traces and therefore it is discussed in a separate Sect. A.2.1.3. This spore is hard to fit in general swimming patterns and can best be described as mixture of *gyration* and *spinning* pattern. This trace is named "PEG-Un-1" in the following.

A.2.1 Exploration on PEG: Swimming Pattern Analysis

The swimming patterns in vicinity of a PEG surface are described with the same motion patterns as used for glass in the last Sect. A.1.1. The name for the trajectories follows the same systematic as used before, e.g. PEG-Gy-1. Only the dark blue trace (PEG-Un-1) visible in Fig. A.22 panel (l) is named differently.

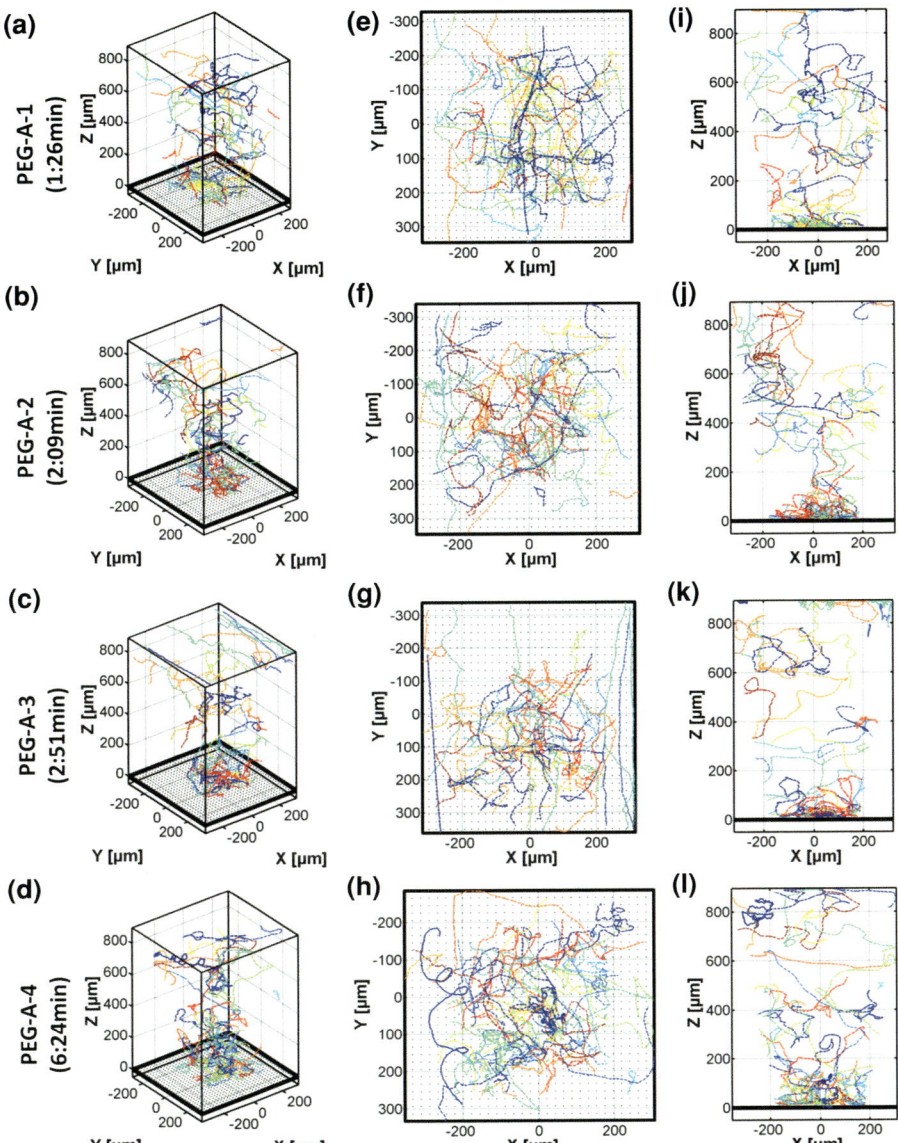

Fig. A.21 3D rendered plots of spores exploring a PEG surface. The spores within a distance of 0–850 μm from the surface are; **a–d** 3D view; **e–h** xy view; **i–l** xz view. To distinguish the trajectories from each other they are colored

All shown trajectories are discussed with the help of five individual plots. In the corresponding Figs. (A.23, A.24, A.25, A.26, A.27, A.28 and A.29) panel (a) shows a 3D rendered plot. This plot gives an overview over the motility. Panel (b) shows the velocity histogram with a fitted Maxwell–Boltzmann distribution [see

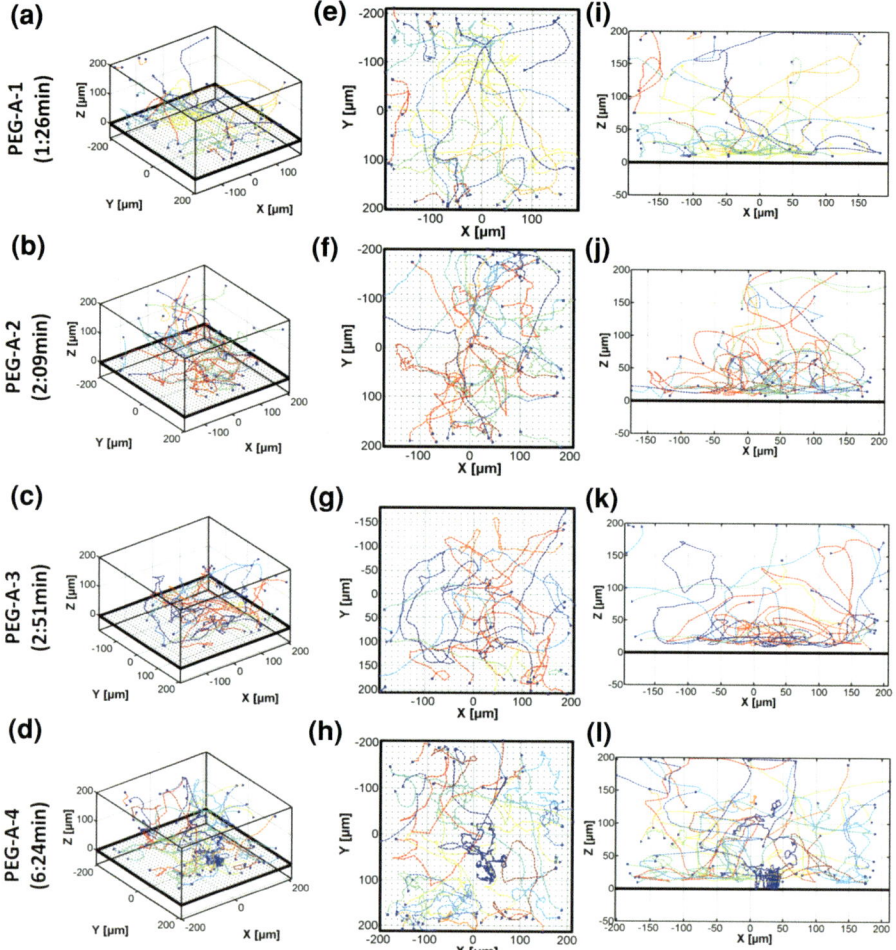

Fig. A.22 3D rendered plots of spores exploring a PEG surface. The spores within a distance of 0–200 μm from the surface are; **a–d** 3D view; **e–h** xy view; **i–l** xz view. To distinguish the trajectories from each other they are colored

Eq. 5.1]. In panels (c–e) detailed exploration parameters are shown and plotted versus the observation time in seconds. These parameters are as follows:

- The change in the Z-position is shown in panels (c, d) in blue and refers to the scale on the right side of the graph.
- The change in velocity is shown in panels (c, e) in black and refers to the scale on the left side of the graph.
- α_v distribution is plotted in red in panels (d, e). Depending whether it is plotted together with the change in the Z-position panels (e) the scale is shown on the left side, or if it is plotted together with the change in velocity panels (e) the scale is plotted on the right side.

Table A.5 Distribution of the determined motion pattern for PEG

Name	Time (min)	Gyration (*hit and run*)	Spinning	Undefined	Total
PEG-A-1	1:26	9 (7)	0	0	16
PEG-A-2	2:09	10 (8)	0	0	18
PEG-A-3	2:51	7 (5)	0	0	12
PEG-A-4	6:24	14 (9)	0	1	24
Sum		40 (29)	0	1	70

In Table A.5 the distribution of the determined motion patterns is summarized.

A.2.1.1 PEG Swimming Pattern: Gyration

In Figs. A.23, A.24 and A.25 three exemplary trajectories for the *gyration* swimming pattern are shown. The spore behavior within this pattern is similar to the previously discussed behavior on AWG (see Sect. A.1.1.1). The timestamp jitter, which is explained in Sect. 4.1.4, is as well present in the data. In fact, it is visible that sometimes the velocity jumps to an extremely high value for a single data point. For example in Fig. A.23, panel (c) the velocity jumps at t $\approx$ 9.5 s from 150 to 760 μm/s and back to 150 μm/s. For the motion analysis velocity values larger than 500 μm/s are therefore ignored.

The swimming pattern of the spore shown in Fig. A.23 is typical for the *gyration* pattern. In the z-projection the spore swims in a wavelike path over the surface. For the first 12 s the spore swims up to a distance of 30–40 μm away from the surface. Towards the end of the recorded trajectory it swims to a distance of 90 μm, where it turns around and swims back towards the surface. It is not possible to identify a correlation between the distance from the surface and a change in α_v [panels (d, e)]. The velocity does not change either with the distance from the surface.

For the spore movement shown in Figs. A.24 and A.25 a similar behavior is observed. The spore swims wavelike over the surface (in Zs). Both traces swim more than 120 μm away from the surface before they turn around and swim again towards the surface. Neither for α_v nor for the swimming speed a correlation between the distance from the surface and changes in α_v or speed is detected. In comparison to the AWG surface less spores are assigned to the *gyration* pattern (see Table A.2 for AWG and Table A.5 for PEG).

The next difference to AWG is that for none of the spores assigned to the *gyration* pattern the determined center of the spore body is observed in the surface plane. The smallest observed distance for the center of mass of the spore body to the surface is 5 μm. Due to the observation that the spores swim along the surface it is extremely likely that they make contact with the surface with their flagella. That the center of mass of the spore body is not observed in the surface plane can be seen in Figs. A.23, A.24 and A.25 and in Figs. A.26, A.27 and A.28 described in the next section.

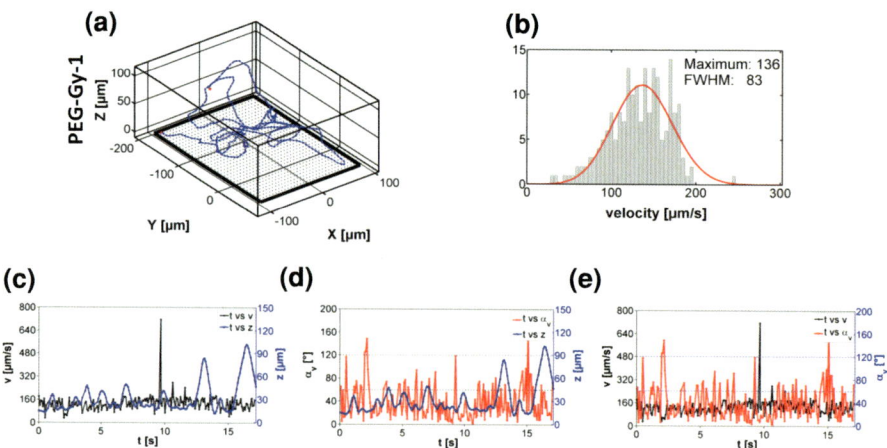

Fig. A.23 Example 1 (PEG-Gy-1) for a movement within the vicinity of the surface **a** 3D rendered trajectory; **b** velocity histogram with fitted Maxwell–Boltzmann distribution; **c** velocity (*black, left side*) and distance to the surface (*blue, right side*) versus elapsed time; **d** α_v (*red, right side*) and distance to the surface (*blue, left*) versus elapsed time; **e** velocity (*black, right side*) and α_v (*red, left*) versus elapsed time

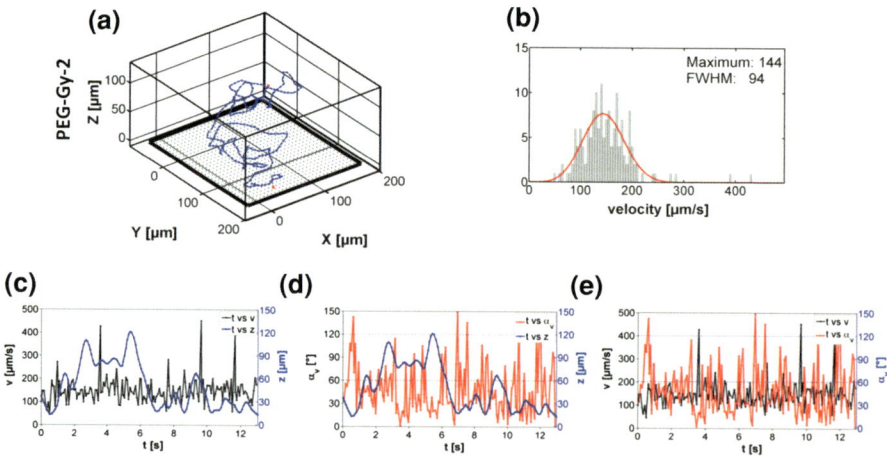

Fig. A.24 Example 2 (PEG-Gy-2) for a movement within the vicinity of the surface. **a** 3D rendered trajectory; **b** velocity histogram with fitted Maxwell–Boltzmann distribution; **c** velocity (*black, left side*) and distance to the surface (*blue, right side*) versus elapsed time; **d** α_v (*red, right side*) and distance to the surface (*blue, left*) versus elapsed time; **e** velocity (*black, right side*) and α_v (*red, left*) versus elapsed time

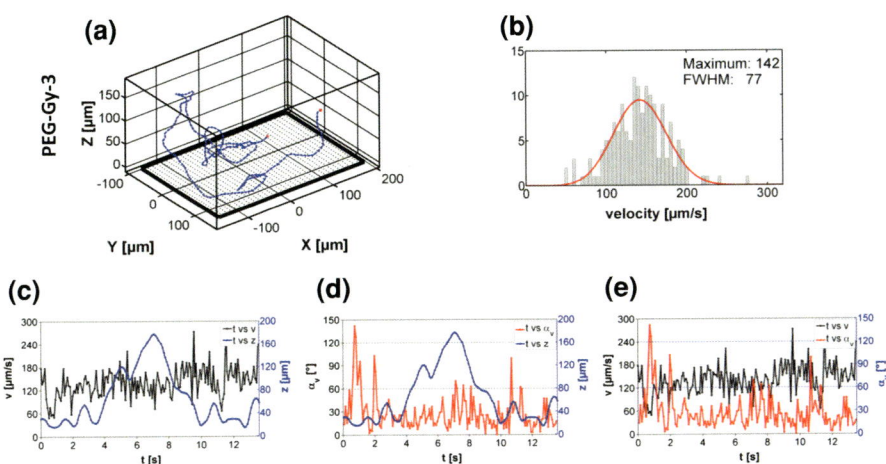

Fig. A.25 Example 3 (PEG-Gy-3) for a movement within the vicinity of the surface. **a** 3D rendered trajectory; **b** velocity histogram with fitted Maxwell–Boltzmann distribution; **c** velocity (*black, left side*) and distance to the surface (*blue, right side*) versus elapsed time; **d** α_v (*red, right side*) and distance to the surface (*blue, left*) versus elapsed time; **e** velocity (*black, right side*) and α_v (*red, left*) versus elapsed time

A.2.1.2 PEG Swimming Pattern: Hit and Run

Figures A.26, A.27 and A.28 show three example trajectories for the *hit and run* swimming pattern. As already pointed out in the description of the motion pattern on AWG, the *hit and run* pattern is a special case of the *gyration* pattern. However, it is listed as an individual pattern because it used to rate for the antifouling performance of the surface. In the hit and away pattern the spores move fast ($v_p > 100$ μm/s) towards and away from the surface. As it is apparent in Figs. A.23, A.24 and A.25, panel (c), the timestamp jitter (see Sect. 4.1.4) occurs in the data.

The spores shown in Figs. A.23, A.24, A.25, A.26, A.27 and A.28 stay close to the surface only for a short time period. For the α_v distribution no clear correlation between the changes in α_v and the distance from the surface is found in the data. The swimming speed does not decrease either when the spore swims close to the surface. In all three trajectories the spore does not get closer than 5 μm away from the surface. All three spores approach the surface from a great distance (between 300 and 400 μm) but do not move towards the surface as straight as the traces recorded on other surfaces. In Table A.5 the spore distribution for the exploration patterns is shown. Many spores can be assigned to the *hit and run* swimming pattern. The population of this pattern is greater than the one for the exploration of AWG (see Table A.2).

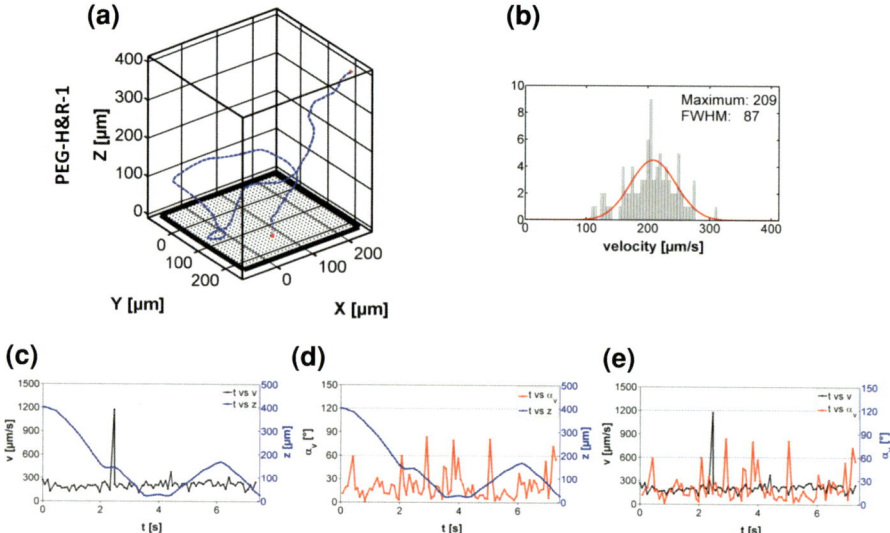

Fig. A.26 Example 1 (PEG-H&R-1) for a *hit and run* movement. **a** 3D rendered trajectory; **b** velocity histogram with fitted Maxwell–Boltzmann distribution; **c** velocity (*black, left side*) and distance to the surface (*blue, right side*) versus elapsed time; **d** α_v (*red, right side*) and distance to the surface (*blue, left*) versus elapsed time; **e** velocity (*black, right side*) and α_v (*red, left*) versus elapsed time

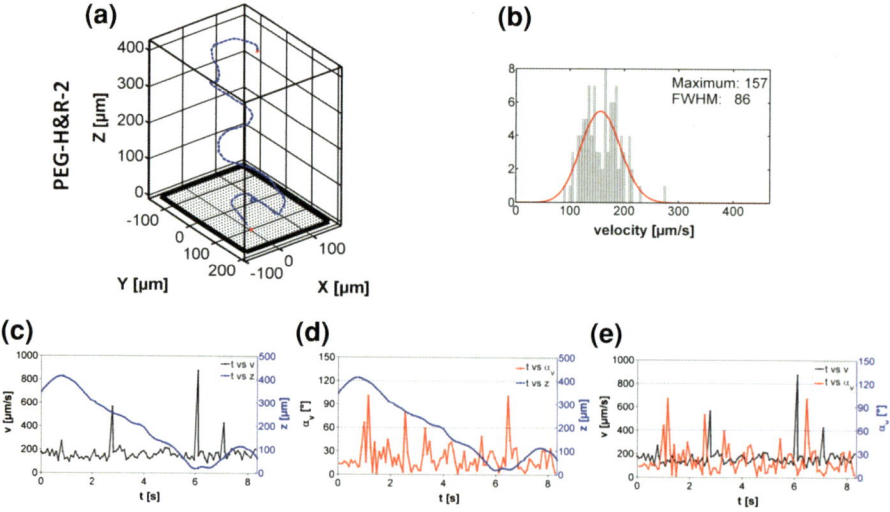

Fig. A.27 Example 2 (PEG-H&R-2) for a *hit and run* movement. **a** 3D rendered trajectory; **b** velocity histogram with fitted Maxwell–Boltzmann distribution; **c** velocity (*black, left side*) and distance to the surface (*blue, right side*) versus elapsed time; **d** α_v (*red, right side*) and distance to the surface (*blue, left*) versus elapsed time; **e** velocity (*black, right side*) and α_v (*red, left*) versus elapsed time

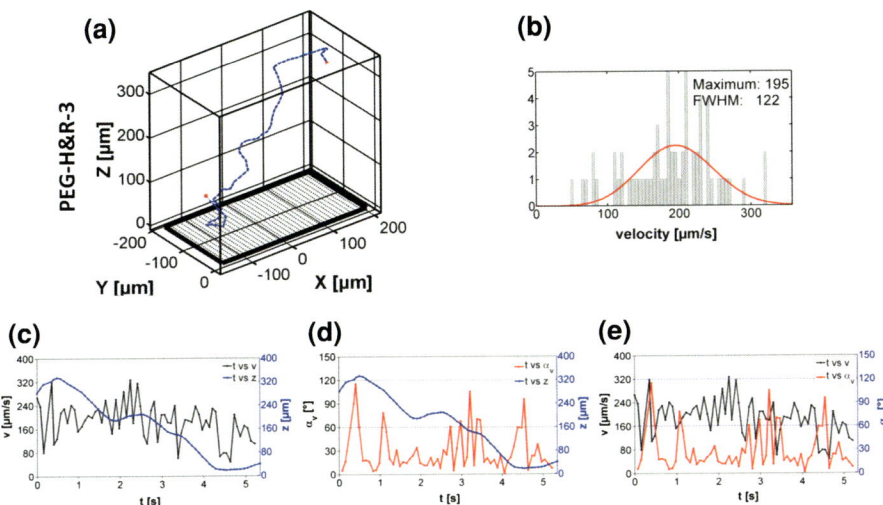

Fig. A.28 Example 3 (PEG-H&R-3) for a *hit and run* movement. **a** 3D rendered trajectory; **b** velocity histogram with fitted Maxwell–Boltzmann distribution; **c** velocity (*black, left side*) and distance to the surface (*blue, right side*) versus elapsed time; **d** α_v (*red, right side*) and distance to the surface (*blue, left*) versus elapsed time; **e** velocity (*black, right side*) and α_v (*red, left*) versus elapsed time

A.2.1.3   Detailed Description of "The Unusual" Spore

Figure A.29 panels (a–e) shows "the unusual" spore of Fig. A.22 panel (l). The spore clearly swims much slower [26 ± 20 μm/s Fig. A.29 panel (b)] than the spores described in the subsection before. It is the only one out of 70 recoded traces exploring a PEG surface where the spore body "touches" the surface. The characteristics of this spore do not fit in the defined swimming pattern. For the *spinning* pattern, the movement in the Z-position is too large and the spore swims in loops over the surface [panels (a, c, d)] rather than staying at the same distance to the surface. Nevertheless, the α_v distribution [panels (d, e)] and the velocity [panels (c, e)] would fit the characteristics of the *spinning* pattern. It also does not fit the *gyration* pattern because the trace is too slow and spends too much time close to the surface. The characteristics of the spore are somewhere between these two swimming patterns. The spore swims mostly at a distance of 25 μm to the surface, but sometimes it swims down to the surface [best seen in Fig. A.29 panels (c, e) blue curve]. As it is discussed in Sect. 4.5 this trajectory is used to determine the position of the surface. The surface position is also checked by individual spores trajectories 30 min later and can be approved at this position.

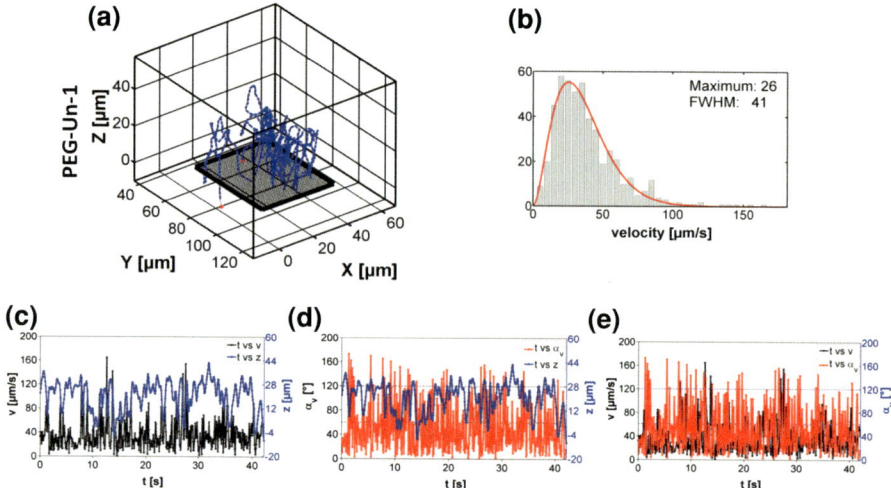

Fig. A.29 Detailed plot for "the unusual" spore (PEG-Un-1). **a** 3D rendered trajectory; **b** velocity histogram with fitted Maxwell–Boltzmann distribution; **c** velocity (*black, left side*) and distance to the surface (*blue, right side*) versus elapsed time; **d** α_v (*red, right side*) and distance to the surface (*blue, left*) versus elapsed time; **e** velocity (*black, right side*) and α_v (*red, left*) versus elapsed time

A.2.2 Exploration on PEG: General Behavior

The following section is organized as the corresponding Sect. A.1.2 for glass. With the help of velocity histograms the spore exploration behavior is analyzed. One aspect is to generalize the results obtained by the analysis of exemplary trajectories in the last section. The results are:

- The spores (except for PEG-Un-1) do not swim up to the surface but rather swim away or along in a distance of 5 μm from the surface.
- The spores are not significantly slower close to the surface than in the bulk.
- No correlation between the α_v distribution and the distance to the surface is found. However it is observed that the spores perform many turns in the area *close* (0–50 μm) and *near* (50–200 μm) the surface.

In Fig. A.30 the velocity histograms are shown for three different sections of the observation volume. The sections are defined in the same way as for AWG (see Sect. A.1.2) and named in the following matter: *bulk* [200–800 μm, panels (a–c)], *near* the surface [50–200 μm, panels (d–f)] and *close* to the surface [0–50 μm, panels (g–i)]. The histograms in Fig. A.30 are also shown for different points in time. The histograms in panels (a, d, g) are recorded after 1:26 min, panels (b, e, h) after 2:51 min and panels (c, f, i) after 6:24 min. The swimming performance in the bulk [panels (a–c)] is already explained in 5.1.1 under the name Bulk-II-A-*.

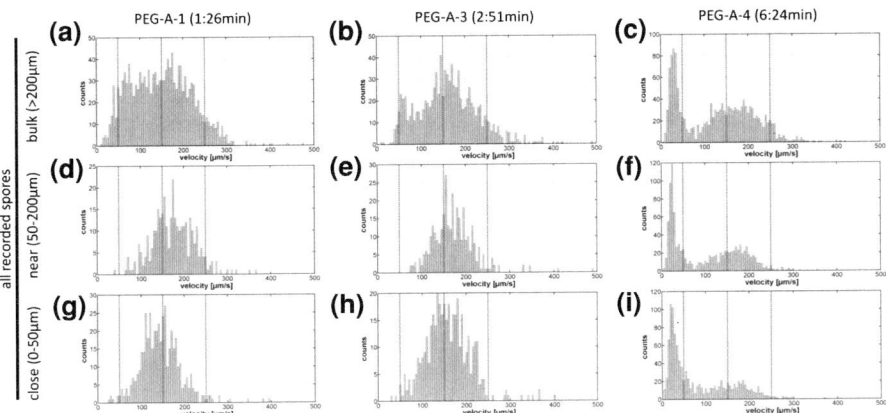

Fig. A.30 Velocity histograms on PEG-A-* shown for three distances **a–c** 200–800 μm, **d–f** 50–200 μm and **g–i** 0–50 μm away from the surface. **a, d, g** is recorded after 1:26 min, b, e, h after 2:51 min and **c, f, i** after 6:24 min. The data shown in **a–c** is already discussed in Sect. 5.1.1 by the name Bulk-II-A-*. The *vertical lines* indicate velocities of 50, 150, 250 μm/s

Fig. A.31 shows the velocity histograms for the same sections and times as Fig. A.30 but only for the spores assigned to the *orientation* pattern [panels (a–c)] and to the *gyration* pattern [panels (d–i)]. The velocity distribution of the unusual trace is also shown in Fig. A.31 [panels (j–l)].

The following analysis is based on Figs. A.30 and A.31. Shortly after the spore injection [1:26 min, panels (a, d, g)] and after the intermediate time [2:51 min, panels (b, e, h)] spores belonging to the *wobbling* pattern are only detected in the *bulk*. At the latest analyzed point in time [6:24 min, panels (c, f, i)] the spores assigned to the *wobbling* pattern are detected in the complete observation volume. This observation is clearly visible for the spores *near* (50–200 μm) the surface [panel (f)] when the histograms for all spores (Fig. A.30) are compared to the histogram for the spores assigned to the *gyration* and *orientation* pattern (Fig. A.31). The situation *close* (0–50 μm) to the surface [panels (i, l)] is more complex. The histogram [Fig. A.30, panel (i)] *close* to the surface is strongly influenced by the occurrence of the "unusual" trajectory (PEG-Un-1, see Sect. A.2.1.3). Looking at the count rate for the slow velocities [Fig. A.30, panel (i)] it becomes clear that the counts of movement vectors observed for "PEG-Un-1" [Fig. A.31, panel (l)] and *gyration* [Fig. A.31, panel (i)] are less than the total number of observed slow velocity vectors. Therefore, spores assigned to the *wobbling* pattern must be present *close* to the surface.

To determine the motility of the spores belonging to the *wobbling* pattern in Fig. A.32 all observed trajectories for PEG-A-4 are shown and are color coded for different velocities. The slow velocity values in the histogram in Fig. A.30, panel (i) are caused by the spore PEG-Un-1 and by the spore named "slow 3". The spore named "slow 3" swims in a typical *wobbling* pattern from a distance of 150 μm to the surface towards the surface. In a distance of 40 μm above the surface it swims

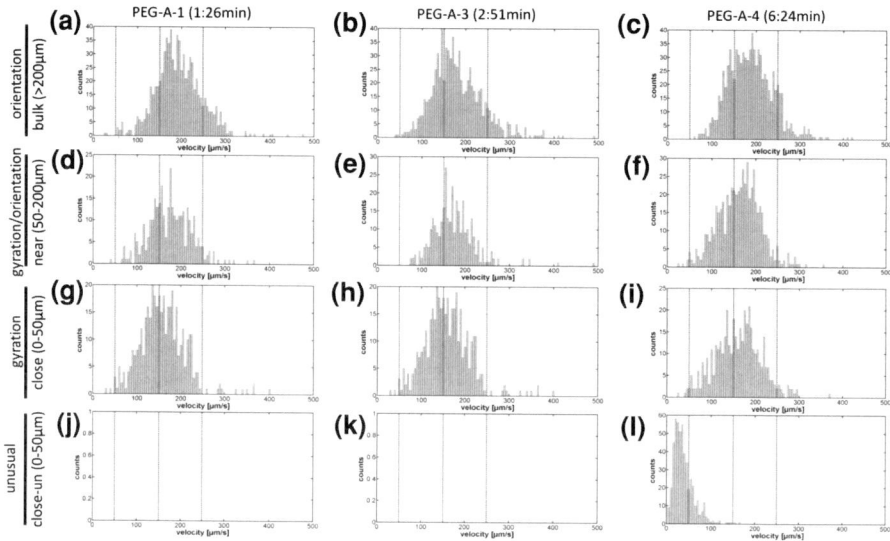

Fig. A.31 Velocity histograms on PEG-A-* for the spores assigned to the *orientation* pattern **a–c**, to *gyration* pattern **d–i** and *PEG-Un-1***j–l**. The histograms are shown for three distances **a–c** 200–800 μm, **d–f** 50–200 μm and **g–l** 0–50 μm from the surface. **a, d, g, j** is analyzed after 1:26 min, **b, e, h, k** after 2:51 min and **c, f, i, l** after 6:24 min. The data shown in **a–c** is already discussed in Sect. 5.1.1 under the name Bulk-II-A-*. The *vertical lines* indicate velocities of 50, 150, 250 μm/s

Fig. A.32 Spore trajectories *close* and *near* the surface (0–200 μm) color coded according to different velocities (*gray*: v < 50 μm/s; *black* 100: < v < 500 μm/s). The slow traces and PEG-Un-1 are marked

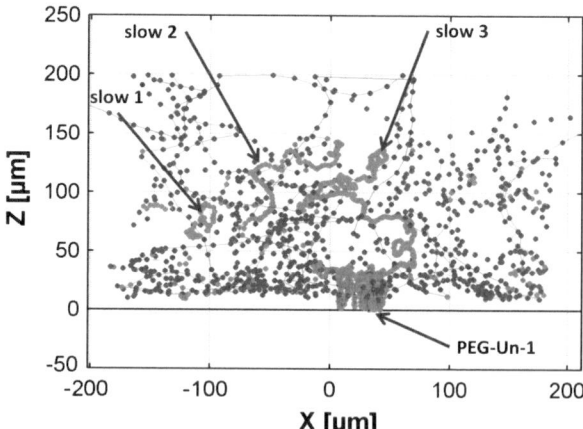

along the surface until it leaves the field of view. The traces "slow 1" and "slow 2" swim parallel to the surface and do neither show a bias to swim towards nor away from the surface. However, these three trajectories ("slow 1–3") explain the big peak for small velocity in the velocities histogram in Fig. A.30, panel (i). In the same histogram the counts for the fast velocities are much smaller than the counts for the slow velocities, even though many fast and individual trajectories

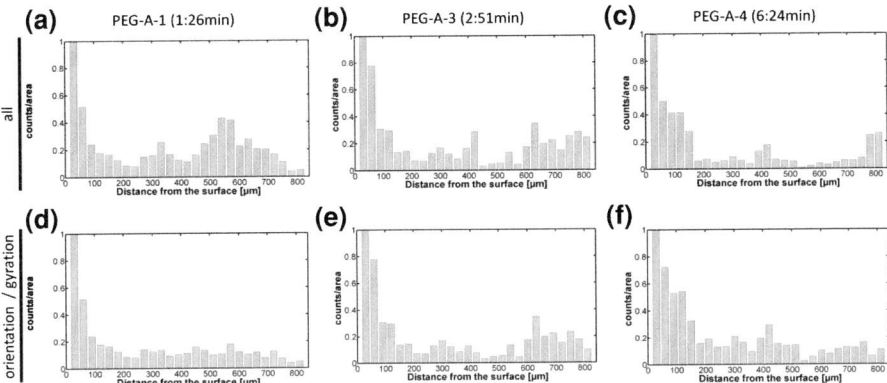

Fig. A.33 General spore distribution on PEG-A-* in the complete observation volume. **a–c** all recorded spores, **d–f** spores only assigned to the *orientation* and *gyration* pattern. **a, d** is recorded after 1:26 min, **b, e** after 2:51 min and **c, f** after 6:24 min

are observed *close* to the surface (see Fig. A.32). This is due to the fact that the majority of trajectories assigned to the *gyration* pattern is in the FoV only for a short time period so that not many counts are present in the histogram shown in Fig. A.30, panel (i). To study and understand the surface exploration the spores assigned to the *gyration* pattern are more important even if the corresponding peak in the histogram is much smaller than the peak of the spores assigned to the *wobbling* pattern.

Figure A.31 shows that the most probable speed (v_p) for the spores belongs to the *orientation* pattern [panels (a–c)] and for the spores assigned to the *gyration* pattern [panels (d–i)] are not significantly different. The spores *close* to the surface swim as fast as the spores *near* the surface or in the bulk. This observation is different to the observation on AWG where it is observed that the velocity *close* to the surface is significantly slower than *near* the surface for the spores assigned to the *gyration* pattern [see Fig. A.14, panels (d–i) and Chap. 6 for a detailed discussion].

In Fig. A.33 the spore distribution in the observation volume is shown. The distribution is shown for all recorded spores in panels (a–c), panels (d–f) show the distributions for the spores assigned to the *orientation* and *gyration* pattern. Even though the spores do not swim completely down to the surface, but rather stay in a distance of 5 μm from the surface, the spore concentration in vicinity to the surface is bigger than in the *bulk*. The spore distribution in the observation volume does not change with elapsing observation time. The spore accumulation is detectable in all shown histograms for the first 200 μm above the surface. Since the spore distribution in the observation volume is fairly constant throughout the complete experiment time, the individual experiments (PEG-A-1 to PEG-A-4) are discussed together for the following analysis.

As for AWG the observation volume is divided into 30 μm slices. The layout of Figs. A.34 and A.35 is the same as for AWG (see Figs. A.16, A.17). In Fig. A.34,

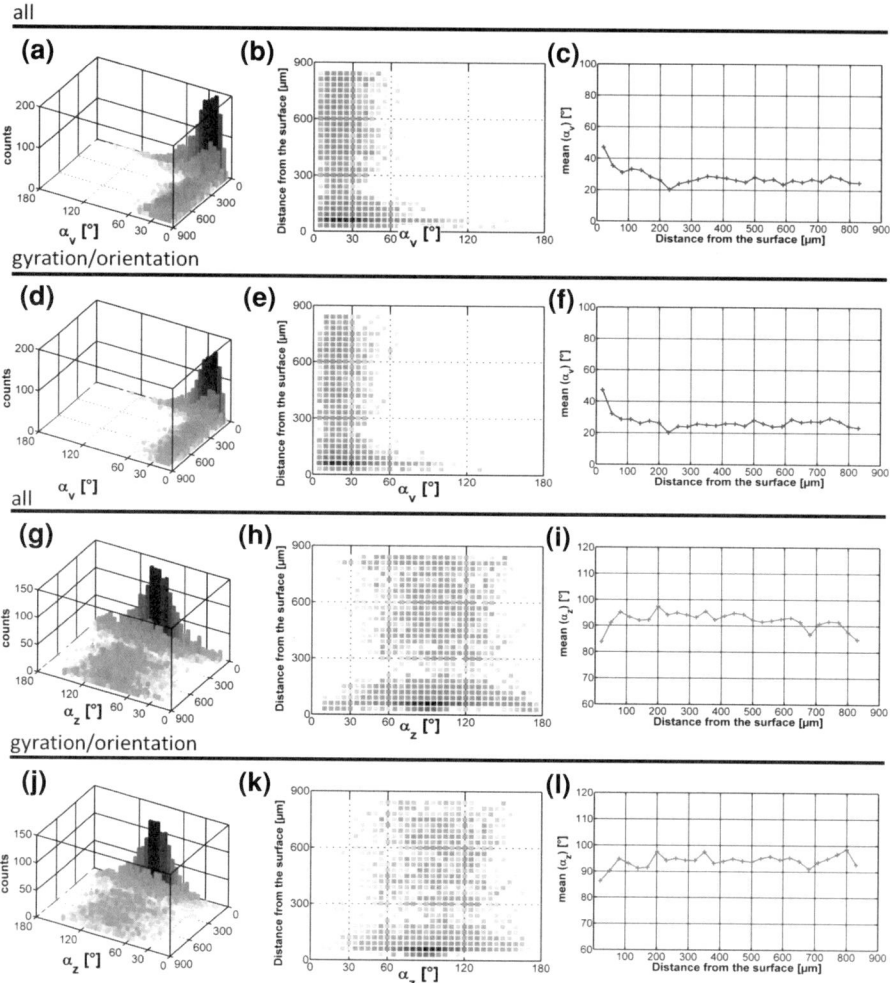

Fig. A.34 α_v and α_z distribution on PEG-A-*. **a, d, g, j** 3D histogram; **b, e, h, k** 2D top view of the perspective 3D histogram; **c, f, i, l** mean value of value shown in the two panels before; **a–c** α_v for all recorded spores; **d–f** α_v for the spores only assigned to the *gyration* pattern, **g–i** α_z for all recorded spores and **j–l** α_z for the spores assigned to the *gyration* pattern. The count rate is encoded in the height of the bars; the darker the bar, the higher the count rate

panels (a–f) the α_v distribution is discussed and the α_z distribution is shown in panels (g–l). In Fig. A.34, panels (a–c) for α_v and panels (g–i) for α_z, the distribution is shown for all spores whereas in Fig. A.34, panels (d–f) for α_v and panels (j–l) for α_z, the spores assigned to the *gyration* and *orientation* pattern are shown.

The α_v distribution *close* to the surface is broader than the one in the *bulk*. The change from the bulk distribution to the surface distribution is observed in the 2D top view of the perspective 3D histogram [panels (b, e)] in a distance of 200 μm

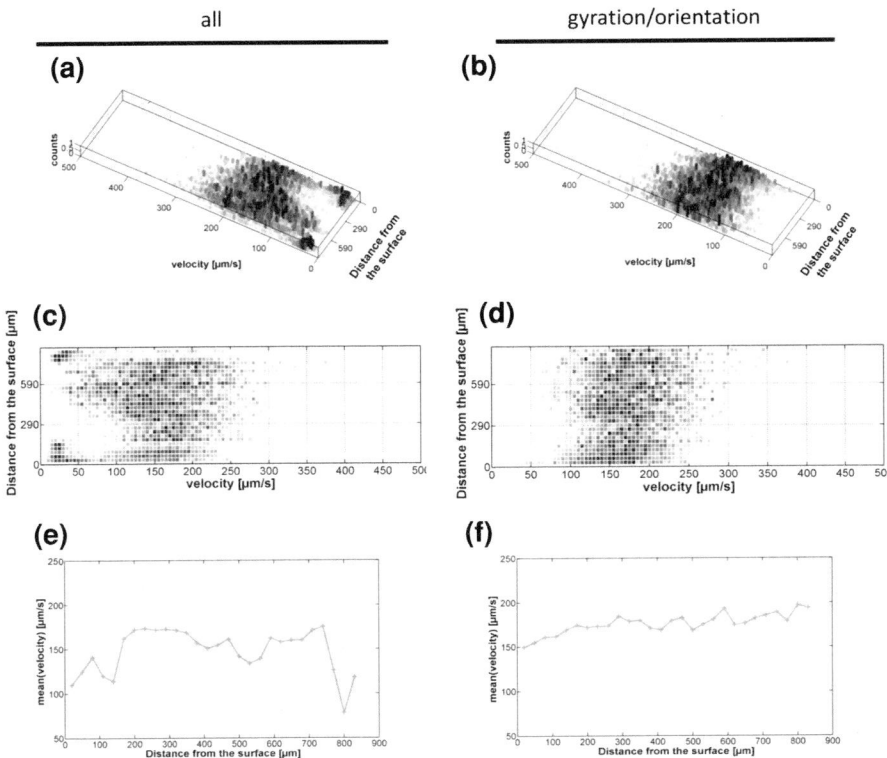

Fig. A.35 Velocity distribution on PEG-A-*. **a, c, e** for all recorded spores; **b, d, f** for the spores assigned to the *gyration* pattern; **a, b** 3D histogram; **c, d** 2D top view of the perspective 3D histogram; **e, f** mean value of the corresponding panel before. The count rate is encoded in the height and in the color of the bars; the darker the bar, the higher the count rate

from the surface. Additionally, for the mean value of α_v [panels (c, f)] the change between the surface distribution and the solution distribution is observed at a distance of 200 μm from the surface. Between all recorded spores and the spores assigned to the *gyration* pattern no significant difference is observed. The distance from the surface where this change is observed coincides with the position of the accumulation of spores in vicinity (0–200 μm) to the surface. A change in the α_v distribution is not expected from the exemplary trace analysis shown in Sect. A.2.1. For the exemplary trace analysis it is not possible to correlate a change in the α_v with the distance to the surface because many changes in the angle occur far away in solution (>100 μm). Therefore, the increase of the α_v distribution can not only be linked to direct surface contact (see discussion AWG) but it describes the *gyration* motion in which the spores swim in a changeful matter towards and away from the surface.

As for α_v distribution, there is no significant difference for α_z distribution between the spore assigned to the *orientation* and *gyration* pattern [panels (j–l)] and to all recorded (g–h) spores. Up to a distance of 60 μm [second data point (panels (i, l))] the spores have a small preference to swim towards the surface.

Figure A.35 shows the velocity distribution in the whole observation volume. In panels (a, c, e) the distribution is shown for all spores and in panels (b, d, f) the velocity distribution is shown for the spores assigned to the *orientation* and *gyration* pattern. For the spores assigned to the *gyration* pattern [panels (b, d, f)] the velocity gets slightly slower when the spores are *near* to the surface. The slowdown is also detected for a distance of 200 μm from the surface. This effect is not seen for all recorded spores [panels (a, c, e)] because the slow spores and the spore "PEG-Un-1" disturb the trend.

In Fig. A.36, the $\bar{\alpha}_v$ distribution [panel (a)] and the mean velocity distribution [panel (b)] are shown in detail for the spores assigned to the *orientation* and *gyration* pattern. The spores' motility is analyzed for three cases: all spores assigned to this analysis (blue), the fraction of spores swimming towards the surface (red) and the fraction of spores swimming away from the surface (green). In the mean velocity distribution in Fig. A.36, panel (b) no significant difference is detected for the fraction of spores swimming towards or away from the surface. In

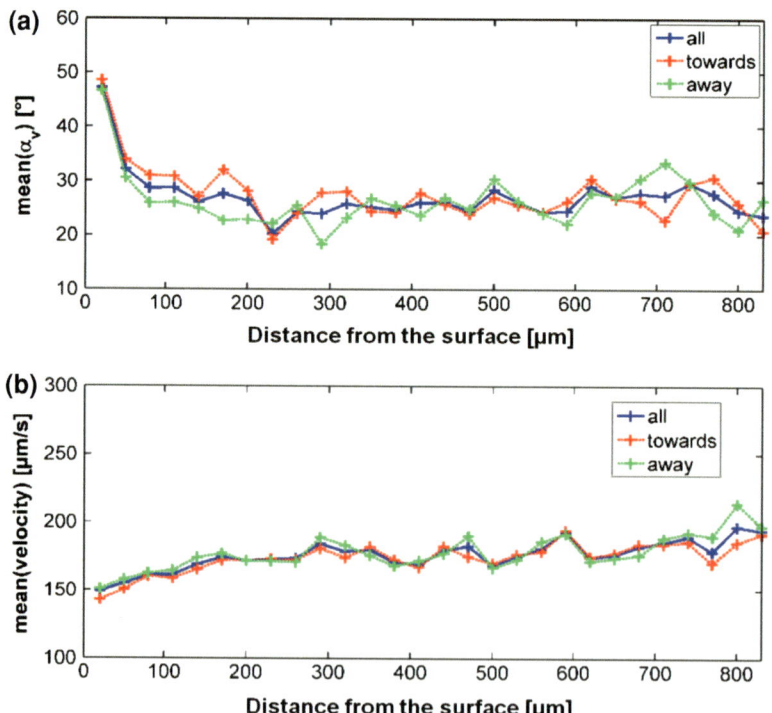

Fig. A.36 Detailed analysis of $\bar{\alpha}_v$ **a** and $\bar{v}$ **b** on PEG-A-*. Only the spores shown assigned to the *gyration* and *orientation* pattern are shown. These spores are analyzed apparently for the spores swimming towards the surface (*red*), all spores (*blue*) and spores swimming away from the surface (*green*)

the $\bar{\alpha}_v$ distribution [panel (a)] for the spore fraction which swims towards the surface (red) and away from the surface (green) in the *bulk* no difference is found, whereas in vicinity (30–150 μm) to the surface a difference is observed. For the spores (assigned to the *gyration* pattern) shown in Fig. A.36, panel (a) which swim towards the surface (red curve) a higher $\bar{\alpha}_v$ value is observed. This means that the spores swimming towards the surface perfore more changes in the swimming direction than the spores swimming away from the surface (green curve). To remember: spores accumulate in vicinity (0–200 μm) to the surface (see Fig. A.33).

In summary, the exploration behavior on PEG can be characterized as follows. The spores do not interact strongly with the surface and except for one individual. The smallest observed distance of the center of mass of the spore body and surface is 5 μm. Furthermore, the surface interaction is small which is verified by the observation that close to the surface the velocity is only slightly slower. This observation leads to the hypothesis that the spore needs a sufficient strong interaction with the surface and the flagella to establish a spore body surface contact. In comparison to AWG the interaction strength between the flagella and the surface appears to be too small support this exploration mechanism. However, the spores still accumulate in vicinity (0–200 μm) to the surface. Approaching the surface the spores perform more turns than while swimming away from the surface.

A.3 Exploration on Fluorinated Monolayer (FOTS) Coating

The last surface discussed is the attractive FOTS surface. In Table A.6 the available statistic is summarized. In total 136 traces and 20,312 data points are analyzed. The experiment in this section is named according to the systematic explained in Sect. 5. The chapter is organized in the same way as the corresponding chapter for AWG (9.1) and PEG (9.2). In Sect. 5.2.3 a detailed settlement analysis is shown based on a direct hologram analysis. Four spores settle in the FoV during this experiment but many more attempt to settle.

Figures A.37 and A.38 panels (a–i) show the spore exploration behavior at three different points in time for all recorded spores over the complete observation

Table A.6 Number of analyzed traces, data points and total observation time, close to the surface

Name	Distance				Elapsed time (min:s)	Duration (s)
	0–50 μm		50–200 μm			
	Number of traces	Number of data points	Number of traces	Number of data points		
FOTS-A-1	13	3,440	17	309	0:29	55.1
FOTS-A-2	33	9,609	32	855	1:24	83.8
FOTS-A-3	72	4,599	47	1,500	5:54	59.4
Sum	118	17,648	96	2,664		198.3

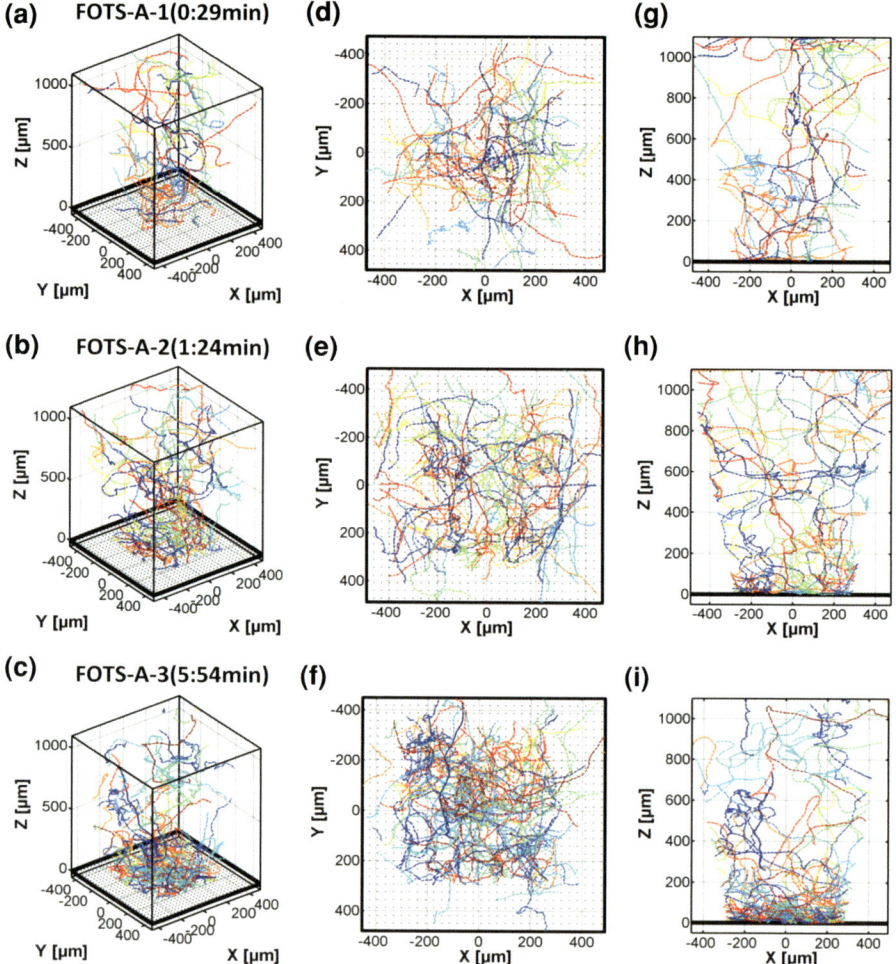

Fig. A.37 Spore trajectories near the surface (0–1100 μm) above the FOTS coating. **a–c** 3D view; **d–f** xy view; **g–i** xz view; To distinguish the trajectories the trajectories are colored differently

volume (Fig. A.37) and for a magnification of the area *near* (0–200) the surface (Fig. A.38). The spores swim independently from each other and in an erratic motion. No swarm behavior or convection is observed in the data. Only for the first point in time data (FOTS-A-1), 29 s after the injection, a preference in the swimming direction of the spores is found. The spores swim towards the surface. This preference exists not only within the first 200 μm from the surface (shown in Fig. A.38) but also in the complete observation volume (Fig. A.37). This preference in the swimming direction is lost with elapsing time. The effect is discussed in detail in Sect. A.3.2.

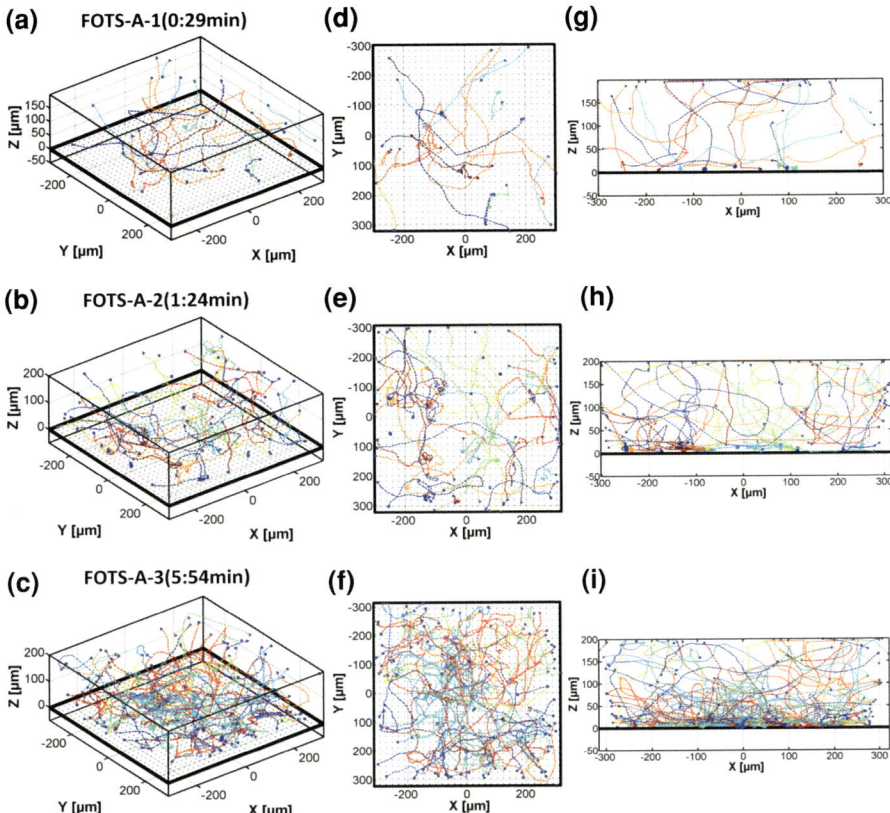

Fig. A.38 Spore trajectories near the surface (0–200 μm) above the FOTS coating. **a–c** 3D view; **d–f** xy view; **g–i** xz view; To distinguish the trajectories the trajectories are colored different

Figure A.38 shows that the exploration behavior changes dramatically with increasing time. If a spore swims close to the surface in the beginning of the experiment [after 29.3 s, panel (g)] it does not explore the surface but rather stick to it. With increasing experiment time this approach pattern vanishes and spores explore the surface in the same way as described for AWG. This is best seen in the xz view of Fig. A.38, panels (g–i) in the change of the spore distribution *close* (0–50 μm) and *near* (50–200 μm) to the surface. The details for the approach patterns are discussed below.

A.3.1 Exploration on FOTS: Swimming Pattern Analysis

In the following the motility patterns are discussed for individual, exemplary traces and is separated in three subsections: *Hit and stick*, *gyration* and *spinning*. *Gyration*

Table A.7 Distribution of the determined motion pattern for FOTS

Name	Time (min)	*Hit and stick*	Gyration (*hit and run*)	Spinning	Settlement	Total
FOTS-A-1	0:29	10	2 (1)	0	0	13
FOTS-A-2	1:24	11	9 (2)	11	0	33
FOTS-A-3	5:54	1	36 (1)	17	1	55
Sum		22	47 (4)	28	1	101

and *spinning* are general motion patterns and are defined in Sect. 5.2.4. *Hit and stick* is a unique motion patter for FOTS. The *hit and run* pattern also occurs but in a smaller percentage than for the other surfaces. Therefore it plays a minor role for the exploration of FOTS surfaces and is not discussed as an individual section. In Table A.7 the occurrence of each pattern is summarized. The trajectories shown in this section are discussed in the same matter as for AWG and PEG. A detailed description can be found in the corresponding Sect. A.1.1 or Sect. A.2.1.

A.3.1.1   FOTS Swimming Pattern: Hit and Stick

The *hit and stick* swimming pattern only occurs on FOTS. The pattern is described with three exemplary traces. Additionally for all analyzed traces the important parameters are summarized in Table A.8.

All traces assigned to this pattern have in common that a spore swims fast ($v_p > 100$ µm/s) and straight towards the surface. Even from a great distance (>90 µm) the approach towards the surface is fairly straight [not many turns, see Figs. A.39, A.40 and A.41, panel (a)]. This visual impression can also be observed for the α_v distribution which is smaller and less changeful ($\alpha_v = 20 \pm 15°$, see Table A.8) on the approach to the surface than in solution ($\alpha_v = 28 \pm 25°$, see Sect. 5.1.3). This straight approach is also observed in the general motion pattern shown in Fig. A.38, panel (c). If a spore reaches the surface it stops swimming immediately [see drop in velocity in Figs. A.39, A.40 and A.41 panel (b), black line]. The spore sticks on the surface for a certain time span before it starts to move according to the *spinning* pattern. Even if the spore detaches from the surface a motion according to the *spinning* pattern always occurs and lasts for thousands or just for a few turns.

The *hit and stick* pattern can be divided into four parts: (i) approach, (ii) sticking to the surface, (iii) *spinning* and (iv$_a$) detachment from the surface or (iv$_b$) settlement. Every part except for the *spinning* part is described in detail in this section. The *spinning* pattern is described as an individual motion pattern in Sect. A.3.1.2. In Figs. A.39, A.40 and A.41 three examples for the *hit and stick* swimming pattern are shown. The motility for the different parts of the *hit and stick* pattern [(i) *approach* panels (a, b), (ii) *sticking* panel (c), (iii) *spinning* panel (d), (iv) detachment (if existing) panels (e, f)] are shown as individual plots.

Figure A.39 shows a spore which enters the FoV 400 µm from the surface. With some turns the spore finds its way towards the surface [see Fig. A.39, panel

Table A.8 Details for trajectories assigned to the *hit and stick* swimming pattern

Number	Frame	v_p (app.) (µm/s)	v_p (det.) (µm/s)	α_z (app.) (°)	α_z (det.) (°)	Sticking (s)	Comment
1	350–1999	221 ± 37	–	151	–	17.1	
2	352–865	160[+]	124[+]	159	70	2.4	Detach
3	442–1999	234 ± 38	–	107	–	18.6	
4	448–794	225 ± 61	204 ± 23	141	62	14.2	Detach
5	504–1999	269 ± 50	–	106	–	21.8	
6	565–999	236 ± 43	–	178	–	15.2	
7	732–1363	203 ± 54	147 ± 36	161	58	31.7	Detach
8	742–1999	251 ± 73	–	168	–	8.3	
9	778–1081	249 ± 64	–	119	79	11.4	Detach
10	797–1088	221 ± 67	150 ± 46	112	57	7.9	Gyration
11	999–1196	short	150 ± 37[‡]	Short	64	10.7	Gyration
12	1028–1187	204 ± 49	160 ± 38[‡]	100	56	3.4	Gyration
13	1037–1999	230 ± 51	120 ± 40[+]	141	–	50.0	
14	1191–1518	236 ± 44	148 ± 56	174	76	11.3	Getach
15	1220–1999	177 ± 29	–	132	–	0.8	Out of gyration
16	1235–1385	242 ± 41[‡]	183 ± 25	166	73	4.3	Getach
17	1276–1999	250 ± 28[‡]	–	155	–	short	Out of gyration
18	1278–1403	185 ± 44[‡]	202 ± 41[‡]	108	86	4.8	Out of gyration[+] gyration
19	1328–1999	245 ± 49	–	132	–	12.8	Out of gyration
20	1468–1637	212 ± 42[‡]	147 ± 37	125	89	1.7	Getach
21	1487–1684	221 ± 49	163 ± 46	170	56	16.0	Detach
Mean		223 ± 48	166 ± 38	140 ± 26	69 ± 12	15 ± 12	

Number	$\bar{\alpha}_v$ (app.) [°]	$\bar{\alpha}_v$ (det.) [°]	af (stick) (°/s)	Dist. app. (µm)	Dist. det. (µm)	Dist. leave (µm)	Comment
1	16.5 ± 13.4	–	378 ± 209	30.1	–	–	
2	32.9 ± 17.7	48.8 ± 25.6	647 ± 249	18.5	6.3	25	Detach
3	23.5 ± 10.8	–	338 ± 213	16.6	–	–	Detach

(continued)

Table A.8 (continued)

Number	Frame	v_p (app.) (μm/s)	v_p (det.) (μm/s)	α_z (app.) (°)	α_z (det.) (°)	Sticking (s)	Comment
4	14.5 ± 10.0	11.5 ± 7.6	417 ± 222	25.0	7.7	85	Detach
5	20.0 ± 17.7	–	283 ± 185	14.1	–	–	
6	24.2 ± 18.8	–	147 ± 109	24.3	–	–	
7	25.5 ± 19.8	28.9 ± 25.0	458 ± 188	23.1	9.5	400	Detach
8	17.1 ± 13.6	–	213 ± 182	20.2	–	–	
9	18.4 ± 11.0	19.5 ± 11.0	325 ± 150	16.4	8.0	200	Detach
10	20.2 ± 21.9	33.6 ± 30.7[†]	203 ± 128	9.0	6.9	24	Gyration
11	short	17.1 ± 7.8[‡]	335 ± 170	Short	12.2	64	Gyration
12	31.3 ± 17.9	40.0 ± 47.5[†]	358 ± 191	8.8	6.2	13	Gyration
13	18.1 ± 16.0	–	411 ± 213	18.3	–	–	
14	13.4 ± 13.5	28.2 ± 18.3	302 ± 183	44.0	8.9	900	Detach
15	17.8 ± 14.7	–	–	12.9	–	–	Out of gyration
16	24.7 ± 16.4[‡]	16.7 ± 9.0	237 ± 108	25.6	7.3	109	Detach
17	10.8 ± 7.6[‡]	–	–	30.6	–	–	Out of gyration
18	18.6 ± 10.8[†]	7.9 ± 4.5[‡]	319 ± 165	15.8	13.4	26	Out of gyration[+] gyration
19	22.8 ± 15.3	–	309 ± 163	10.9	–	–	
20	17.5 ± 12.3[‡]	12.5 ± 9.1	515 ± 339	6.9	5.7	80	Out of gyration
21	27.9 ± 23.7	17.6 ± 12.7	788 ± 261	23.5	4.4	152	Detach
Mean	20 ± 15	18 ± 12	367 ± 191	20 ± 9	8 ± 3		

app Approach to the surface; *det* Detach from the surface; *dist* Distance from the surface (for an explanation of "[+], [‡], [†], short" please refer to the description in the text)

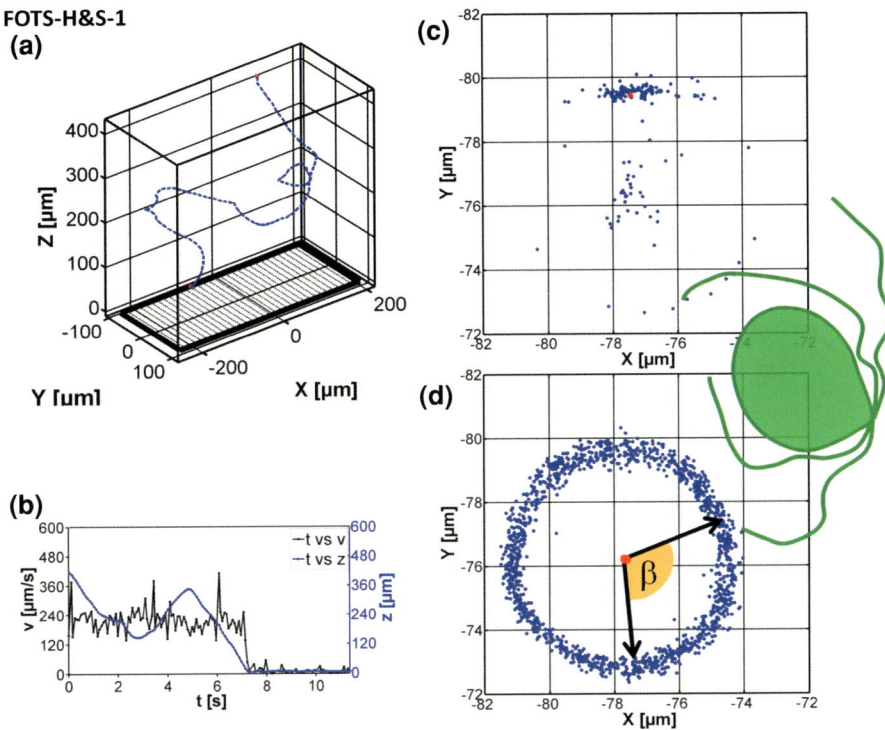

Fig. A.39 Example 1 (FOTS-H&S-1) for a *hit and stick* movement. **a** 3D rendered trajectory for the approach to the surface; **b** velocity (*black, left side*) and distance to the surface (*blue, right side*) versus elapsed time for the approach and the beginning of the *sticking* phase; **c** xy view of the sticking phase; **d** xy view of the spinning phase. A sketch of the spore is shown to clarify the dimension of the *sticking* and *spinning* phase. The spinning phase is characterized by the angle β and the radius (ra) as defined in Fig. 4.18

(a)]. On the approach to the surface the spore swims at an average velocity of 212 ± 37 μm/s. On this path towards the surface the spore swims slightly more directed (with less changes) [$\alpha_v = 16 \pm 13°(\pm 80\%)$] than it is typically observed for a spore assigned to the *orientation* pattern [$\alpha_v = 28 \pm 25°(\pm 90\%)$, see Sect. 5.1.3]. After the spore reaches the surface it stops swimming immediately [see panel (b), black line] and stays at the same position until the end of the recording (143 s). During that time the spore sticks for about 17.1 s before it starts moving according to the *spinning* pattern. The motion of the sticking phase is shown in [panel (c)]. During this phase the spore position changes by some infrequently "flips" of the spore on the surface. Subsequently to the sticking the spore starts spinning (rf $= 187 \pm 30$ rpm; ra $= 3.6 \pm 0.2$ μm and af $= 1124 \pm 179°$/s) until the end of the recording. This spore neither leaves the surface nor settles during the observation time. Based on the analysis in FOTS-A-3 it is known that the spore left the surface at some point in time between FOTS-A-2 and FOTS-A-3.

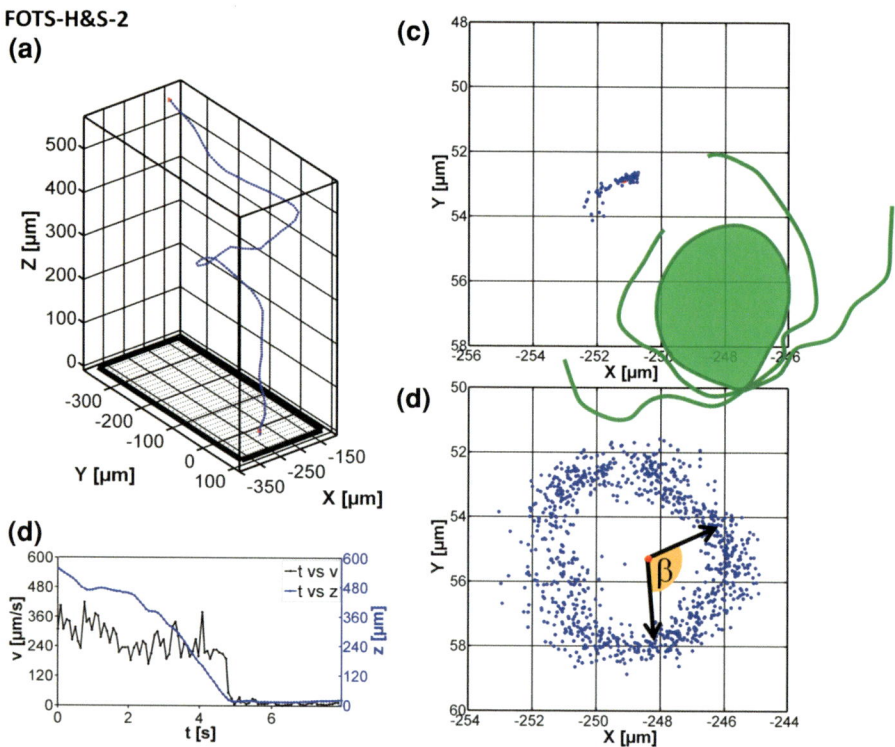

Fig. A.40 Example 2 (FOTS-H&S-2) for a *hit and stick* movement. **a** 3D rendered trajectory for the approach to the surface; **b** velocity (*black, left side*) and distance to the surface (*blue, right side*) versus elapsed time for the approach and the beginning of the *sticking* phase; **c** xy view of the sticking phase; **d** xy view of the spinning phase. A sketch of the spore is shown to clarify the dimension of the *sticking* and *spinning* phase. The spinning phase is characterized by the angle β and the radius (ra) as defined in Fig. 4.18

The spore shown in Fig. A.40 has the same characteristics in its swimming performance. The spore approaches the surface with a velocity of 251 ± 73 μm/s. The sticking phase last 8.3 s and the *spinning* characteristics are rf = 152 ± 30 rpm; ra = 3.5 ± 0.4 μm and af = 914 ± 179°/s.

The spore shown in Fig. A.41 approaches the surface in the same way as the already discussed spore trajectories except that this spore is able to leave the surface again. The shown detachment from the surface occurs after an extremely short *spinning* period. It was possible to follow the approach part of the trajectory for more than 800 μm towards the surface which is unusual because normally the spore leaves the FoV already after a shorter travel distance. In the approach all the obtained parameters are typical for a fast moving spore: v_p is 248.9 ± 60.5 μm/s and α_v is mostly below 30° but never larger than 50° (not shown as an additional figure). To swim this long distance the spore needs 7.4 s. Subsequently it sticks [panel (c)] to the surface for 11.4 s before it starts to spin for a few turns [panel (d)]. Afterwards it detaches from the surface [panel (e, f)]. The spore spins for a

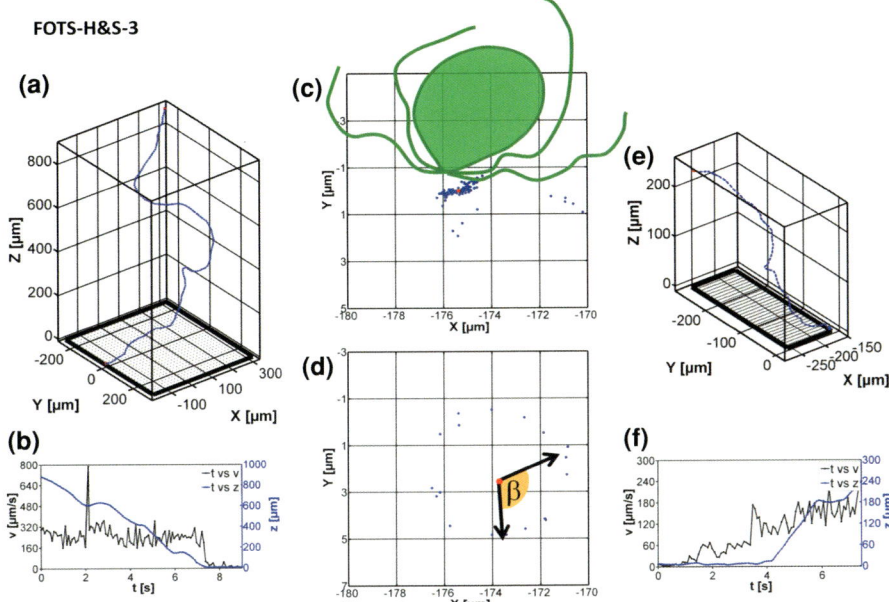

Fig. A.41 Example 3 (FOTS-H&S-3) for a *hit and stick* movement. **a** 3D rendered trajectory for the approach to the surface; **b** velocity (*black, left side*) and distance to the surface (*blue, right side*) versus elapsed time for the approach and the beginning of the *sticking* phase; **c** xy view of the sticking phase; **d** xy view of the spinning phase. A sketch of the spore is shown to clarify the dimension of the *sticking* and *spinning* phase; **e** 3D trajectory for the detachment from the surface; **f** velocity (*black, left side*) and distance to the surface (*blue, right side*) versus elapsed time for the detachment from the surface. The spinning phase is characterized by the angle β and the radius (ra) as defined in Fig. 4.18

short period—only 2.2 s [see panel (d)]—before it leaves the surface, but the characteristics are typical for the spinning phase (rf = 191 ± 27 rpm; ra = 2.9 ± 0.1 μm; af = 1147 ± 164). The first phase of the detachment from the surface is visible in Fig. A.41, panel (f). Suddenly the velocity increases and α_v gets smaller (40°, not shown as an individual figure). The spore leaves the surface with an angle (α_z) of 79° towards the surface normal. That means that the spore swims parallel to the surface for a short time period before swimming back into the solution. The final detachment phase is shown in panel (e). The spore leaves the FoV at a distance of 200 μm from the surface.

In the section of trace "FOTS-H&S-3", shown in Fig. A.41, the spore swims towards the surface at a speed (v_p) of 249 ± 64 μm/s. After surface contact it swims at a velocity (v_p) of 150 ± 46 μm/s which is slower than it was before surface contact. From a closer look at the velocity distribution in panel (f) one can see that the velocity increases with increasing time of the spore swimming in solution. It is not known whether the swimming speed gets as fast as it was before the surface contact because the spore leaves the FoV before a maximum velocity is reached.

In Table A.8 important parameters for the *hit and stick* swimming pattern are summarized for the spores observed in FOTS-A-1 (0:29 min) and FOTS-A-2 (1:24 min). In FOTS-A-3 (5:54) the *hit and stick* swimming pattern is not observed any more (see Table A.7). In Table A.8 spores are listed in the order of their occurrence during the recording. For all observed spores the following parameters are shown (in brackets the abbreviations used in Table A.8 are defined):

- Number of the trace (number)
- First frame and last frame of the observed spore (frame)
- Approach velocity (v_p app.)
- Detachment velocity (v_p det.)
- α_z for the approach to the surface (α_z app.)
- α_z for the detachment from the surface (α_z det.)
- Time the spore sticks to the surface (sticking)
- A comment for the behavior (comment)
- The spore swims further in the *gyration* pattern *(gyration)*
- A part of the trajectory is assigned to the *gyration* pattern before the spore movement is classified to the *hit and stick* pattern (out of *gyration*)
- Spore leaves the surface (detach)
- The mean value of α_v for the approach to the surface ($\bar{\alpha}_v$ app.)
- The mean value of α_v for the swimming performance after the detachment from the surface ($\bar{\alpha}_v$ det.)
- The mean value of angular frequency (af, $\frac{\beta}{dt}$) for the sticking phase (af stick)
- The distance from the surface for the approach when the velocity gets slower (dist. app.)
- The distance from the surface for the detachment from the surface when the velocity increases (dist. det.)
- The distance from the surface when the spore leaves the FoV (dist. leave)
- Some phases (approach, sticking, *spinning*, detach) of the pattern can be extremely short. For these cases it is not feasible to determine a mean value (e.g. spore 19 spins on the surface but the pattern is too short to determine a mean *spinning* speed) (short)
- If a phase (approach, sticking, *spinning*, detach) of the *hit and stick* pattern is short (but bigger as the case "short", see above), a value (e.g. velocity) for this part can be estimated. These values are not as reliable as the other obtained values and therefore no error is stated. For example trace number 2 wriggles/ spins close at the edge of the FoV. Therefore the approach and the detachment are extremely short but observable ($—^+$)
- The spore swims in the *gyration* pattern but for the determination of v_p or α_v a part of the *gyration* pattern is chosen were the spore swims above the solution ($—^{\ddagger}$)
- The spore swims in the *gyration* pattern but is not possible to determine a value for v_p or α_v for a section without surface contact ($—^{\dagger}$)

From the data in Table A.8 the following conclusions can be drawn.

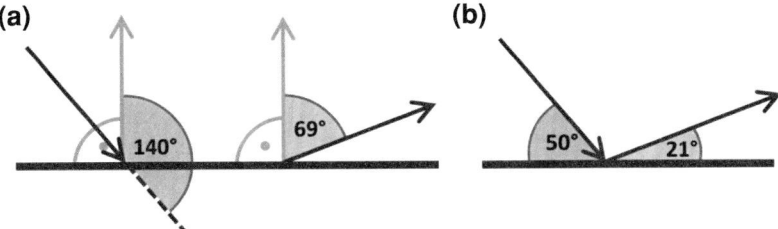

Fig. A.42 Sketch for the approach angle (*green*) and detachment angle (*orange*) observed for spores assigned to the *hit and stick* pattern. **a** shows the determined α_z (angel against the surface normal); **b** for a better understanding the same angle is drawn within respect to the surface plane

- Spores swim faster towards the surface than they swim after they have been *spinning* on the surface. The mean velocity for the approach is 223 ± 48 μm/s and for the detach 161 ± 38 μm/s. It was possible to follow "spore 14" after the detachment from the surface to a distance of 900 μm from the surface. This spore still swims slower after the detachment than during the approach (app.: 236 ± 44 μm/s, det.: 148 ± 56 μm/s). The mean angle ($\bar{\alpha}_v$) distribution is not significantly different for the approach ($20 \pm 15°$) and for the detachment ($18 \pm 12°$) from the surface.
- Spores approach the surface steeper than they detach from the surface (see Fig. A.42 or Table A.8).
- With elapsing time (see frame number) the *hit and stick* pattern changes. In the beginning no spore is able to explore the surface. All spores which come very close to the surface stick to the surface and subsequently start to spin. The spores 15, 17, 18 and 20 show the typical *gyration* pattern. They swim down to the surface, explore it for a certain time (no sticking or *spinning*) and swim back into the vicinity (0–200 μm) of the surface. After the *gyration* part "spore 20" sticks to the surface (short *spinning* phase) before it leaves the surface and the FoV at a distance of 80 μm away from the surface. The "spores 15" and "17" first "explore" the surface (*gyration*) before they stick to the surface at a different position for an extremely short time. Both spores start to spin and do not leave the FoV before the end of the recording. The recorded trajectory of "spore 18" can be split into three parts: *gyration*, *hit and stick* and again *gyration*. With elapsing time the sticking phase gets shorter until it vanishes. From then onwards most spores explore the surface via the *gyration* pattern. In the last analyzed time frame (FOTS-A-3) the *hit and stick* behavior is nearly lost completely (see Table A.7).
- While approaching the surface the velocity drops at a distance of 20 ± 9 μm from the surface. If a spore detaches from the surface the velocity increases at a distance of 8 ± 3 μm.

A.3.1.2   FOTS Swimming Pattern: Spinning

Prior commitment to settlement the spore spins above the surface for a various amount of time. This connection (*spinning*–settlement) has been already depicted

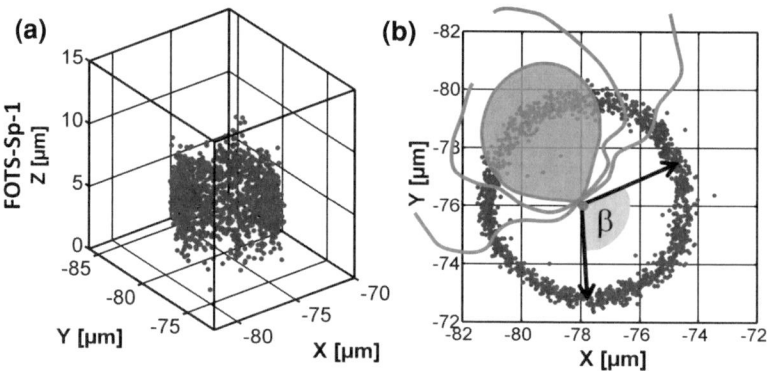

Fig. A.43 Example 1 (FOTS-Sp-1) for *spinning* on FOTS. **a** 3D rendered view; **b** xy view. A sketch of the spore is shown to clarify the dimension of the spore motion. The spinning phase is characterized by the angle β and the radius (ra) as defined in Fig. 4.18

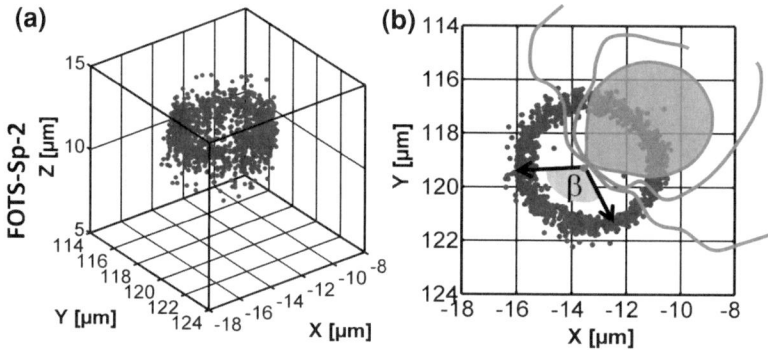

Fig. A.44 Example 2 (FOTS-Sp-2) for *spinning* on FOTS coating: **a** 3D rendered view; **b** xy view. A sketch of the spore is shown to clarify the dimension of the spore motion. The spinning phase is characterized by the angle β and the radius (ra) as defined in Fig. 4.18

in an earlier study [2]. Settlement is an irreversible step in the spore lifecycle and it determines were the spore starts to grow into a new plant. To be able to design an antifouling coating it is important to study the spinning behavior in detail to understand the motion which leads to settlement.

In Figs. A.43 and A.44 the details of the *spinning* pattern are shown. Figure A.43, panels (a, b) shows the 3D rendered and the xy view of the *spinning* pattern. The spore continues *spinning* until the end of the recording. The center of mass of the spore changes on a circle with a diameter of 3.6 $\pm$ 0.1 μm. The angular frequency (af; $\frac{\beta}{dt}$) is 1124 $\pm$ 179°/s. Depending on these values the rotation frequency (rf) can be calculated, which is rf = 187 $\pm$ 29 rpm. The rotation frequency is in a good agreement with the rotation frequency (240 rpm) determined before [2].

Table A.9 Details for the *spinning* spores assigned to the *hit and stick* swimming pattern

Number	Frame	rf (spin) (rpm)	ra (μm)	af (spin) (°/s)
1	350–1999	187 ± 29	3.6 ± 0.1	1124 ± 179
2	352–865	206 ± 46	2.5 ± 0.1	1238 ± 277
3	442–1999	230 ± 33	2.6 ± 0.1	1383 ± 199
4	448–794	168 ± 39	3.0 ± 0.2	1008 ± 234
5	504–1999	219 ± 33	3.1 ± 0.2	1314 ± 198
6	565–999	209 ± 34	2.5 ± 0.2	1259 ± 206
7	732–1363	Short	Short	Short
8	742–1999	152 ± 29	3.4 ± 0.4	914. ± 178
9	778–1081	191 ± 27	2.8 ± 0.0	1147 ± 164
10	797–1088	148 ± 23	3.0 ± 0.2	890 ± 143
11	999–1196	195 ± 39	3.3 ± 0.2	1170 ± 239
12	1028–1187	Short	Short	Short
13	1037–1999	205 ± 35	3.2 ± 0.1	1233 ± 210
14	1191–1518	Short	Short	Short
15	1220–1999	167 ± 39	3.3 ± 0.4	1004 ± 234
16	1235–1385	187 ± 25	2.8 ± 0.5	1124 ± 152
17	1276–1999	217 ± 30	3.2 ± 0.1	1302 ± 183
18	1278–1403	238 ± 45	3.7 ± 0.3	1432 ± 273
19	1328–1999	206 ± 35	2.6 ± 0.2	1239 ± 211
20	1468–1637	156 ± 38	3.2 ± 0.2	938. ± 232
21	1487–1684	Short	Short	Short
Mean		193 ± 27	3.1 ± 0.4	1160 ± 164

spin Spinning; *shorts* The *spinning* phase is too short to determine a mean value for rf, ra and af

The rotation frequency analyzed in FOTS-A-1 and FOTS-A-2 is constant even if a spore spins for several minutes. In experiment FOTS-A-3—recorded 4:30 min later than FOTS-A-2—most of the *spinning* spores of the experiment FOTS-A-1 and FOTS-A-2 have left the surface, only two spores still spin. The rotation frequency is not significantly slower than before. In Table A.9 the mean rotation frequency, the mean radius (ra) and angular frequency (af) is shown. The mean rotation frequency is 193 ± 27 rpm, the radius is 3.1 ± 0.4 μm and the mean angular frequency is 1160 ± 164°/s. The high mean angular frequency in combination with the small standard deviation is a definite indicator for a *spinning* motion. During the sticking phase, in comparison, the value of the mean angular frequency is significantly smaller with a bigger relative variation (367 ± 191°/s, see Table A.8).

Figure A.44 shows another example of the *spinning* pattern. The characteristics (rf = 230 ± 33 rpm, ra = 2.6 ± 0.1 μm, af = 1383 ± 199°/s) for this spore are similar to the one previously discussed.

In experiment FOTS-A-3 one spore out of the recorded 17 *spinning* spores stops *spinning* and settles in the FoV. This spore does not belong to the two spores which have been on the surface already in FOTS-A-1 or FOTS-A-2. Since it is already on the surface in the first analyzed frame of FOTS-A-3 it must have come

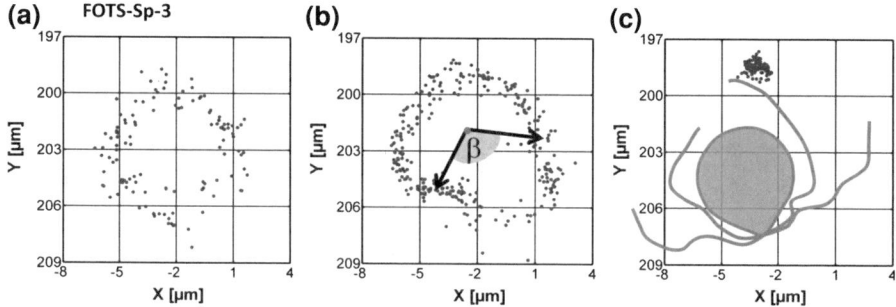

Fig. A.45 Example 3 (FOTS-Sp-3) for a *spinning* spore on the FOTS coating which stops *spinning* and settles in the FoV. **a** xy-view of the *spinning* part I; **b** xy-view of the *spinning* part II; **c** xy-view of the settlement. A sketch of the spore is shown to clarify the dimension of the spore motion. The spinning phase is characterized by the angle β and the radius (ra) as defined in Fig. 4.18

to the surface at any point between FOTS-A-2 and FOTS-A-3. Therefore, it is not known how long the spore actually spins on the surface. The trajectory including the settlement event is shown in Fig. A.45 and can be divided into three parts: *spinning* part I [panel (a)], *spinning* part II [panel (b)] and settlement.

For *spinning* part I [panel (a)] the value of the rotation frequency (122 ± 35 rpm) is already slower than it is observed for the other spinning patterns (193 ± 27 rpm). Consequently, the values of other *spinning* parameters are also slightly different in comparison to the typically observed values listed in Table A.9. The radius (ra = 4.5 ± 0.8 µm) is slightly bigger and the angular velocity is detectable slower ($737 \pm 213°$/s). For the *spinning* part II the spore does not rotate anymore because the value of the rotation frequency for an assumed rotation is very slow (rf = 50 ± 20 rpm). The motion is better described by a twitching motion where the cell body flips around on the surface. The radius in which the center of mass is distributed is bigger and more changeful (ra = 7.0 ± 1.8 µm) than observed for the typical spinning motion. The angular frequency (af = $300 \pm 124°$/s) is also slower meaning that the spore does not move as fast as during the spinning. In the last part [settlement, panel (c)] the spore finally stops moving and is adhered to the surface. This spore settles on the surface and establishes the link between *spinning* and settlement.

A.3.1.3 FOTS Swimming Pattern: Gyration

With elapsing time the *hit and stick* motion pattern vanishes and *gyration* pattern occurs. In FOTS-A-2 the change in the exploration behavior is observed. After a couple of minutes the spores are not longer immediately trapped when they approach the surface but are able to explore the surface. In the FOTS-A-3 experiment *spinning*, *gyration* and *hit and run* patterns as well as a settlement are observed. For the *gyration* motion on FOTS a typical example is shown in

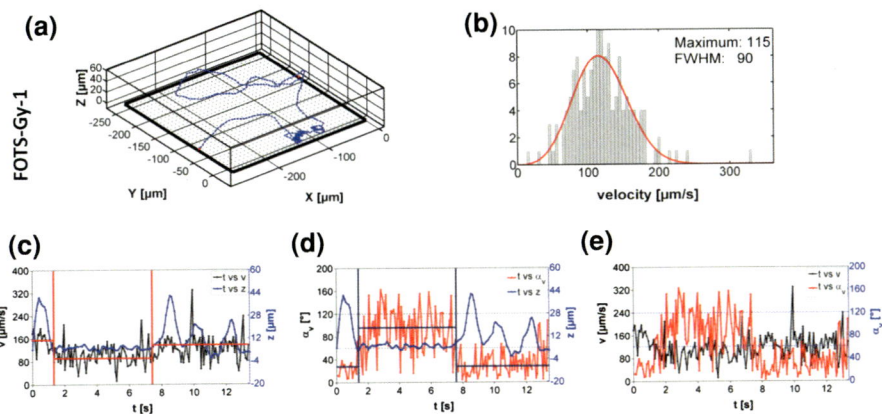

Fig. A.46 Example 1 (FOTS-Gy-1) for a movement within the vicinity of the surface. **a** 3D rendered trajectory; **b** velocity histogram with fitted Maxwell–Boltzmann distribution; **c** velocity (*black, left side*) and distance to the surface (blue, right side) versus elapsed time; **d** α_v (*red, right side*) and distance to the surface (*blue, left*) versus elapsed time; **e** velocity (*black, right side*) and α_v (*red, left*) versus elapsed time

Fig. A.46. The spore swims towards the surface, "examines" the surface, swims away and starts to "explore" the surface at a different position. As already stated for AWG (Sect. A.1.1), spores are slower when they are close to the surface. In Fig. A.46, panels (c, e) this correlation is also shown for the *gyration* pattern on FOTS. The α_v distribution changes as well with the distance to the surface [see panels (d, e)] coinciding with the velocity. For the approach to the surface the spore swims slower at a distance of 17.3 μm from the surface. In a distance of 8.8 μm from the surface the spore detaches and starts swimming faster again. These observed values for the example shown in Fig. A.46 are in the same range as the values determined for the *hit and stick* pattern summarized in Table A.8.

A.3.2   Exploration on FOTS: General Behavior

The exploration behavior on FOTS is complex and changes with elapsing time. In the following the observations of the last section are put in a general context. The focus will be to determine the surface interaction distance and to verify the change in the exploration behavior. In the last section the following observations were described.

• In the beginning of the experiment the *hit and stick* pattern occurs. Since, nearly all observed trajectories in FOTS-A-1 can be assigned to the *hit and stick* pattern the spores are trapped at the surface at a specific position (*sticking* and *spinning* phase) and do not explore the surface as observed on AWG and PEG (occurrence of the *gyration* pattern on these surfaces).

- After 5:54 min the *hit and stick* pattern vanishes and the surface is explored according to the *gyration* pattern. The exploration behavior at this time is similar to the behavior on AWG (see discussion in Sect. 6.3).
- The first permanent settlement event is not witnessed before FOTS-A-3 when the hit and stick pattern has vanished completely.
- For a spore assigned to the *gyration* pattern the α_v distribution increases significantly and the speed is significantly lower if the spore is close to the surface (the same is observed on AWG for the spores assigned to the *gyration* pattern, but is not observed on PEG).

To reduce the complexity of the analysis, first the velocity histograms in Figs. A.47 and A.48 are discussed for three sections of the observation volume: *bulk* [panels (a–c), 200–800 µm], *near* the surface [panels (d–f), 50–200 µm] and *close* to the surface [panels (g–l), 0–50 µm]. The histograms are shown for three different analyzed times: FOTS-A-1 [panels (a, d, g) 0:29 min], FOTS-A-2 [panels (b, e, h) 1:24 min] and FOTS-A-3 [panels (c, f, i) 5:54 min]. The distribution in the bulk is already discussed in Sect. 5.1.1 under the name Bulk-III-A-*.

In Fig. A.47 the velocity histograms are shown for all analyzed spores whereas in Fig. A.48 histograms are shown for the *orientation* pattern [panels (a–c)], for the spores assigned to the *gyration* pattern [panels (d–i)] and for the spores assigned to the *hit and stick* and *spinning* pattern [panels (j–l)].

The following discussion is based on Figs. A.47 and A.48. Shortly after the injection (FOTS-A-1) no spores assigned to the *wobbling* pattern are detected in the complete observation volume. However, the two clearly distinguishable spore fractions observed *close* (0–50 µm) to the surface [panels (g)] are assigned to the *hit and stick* pattern [see Fig. A.48, panel (j)]. This spore velocity distribution is

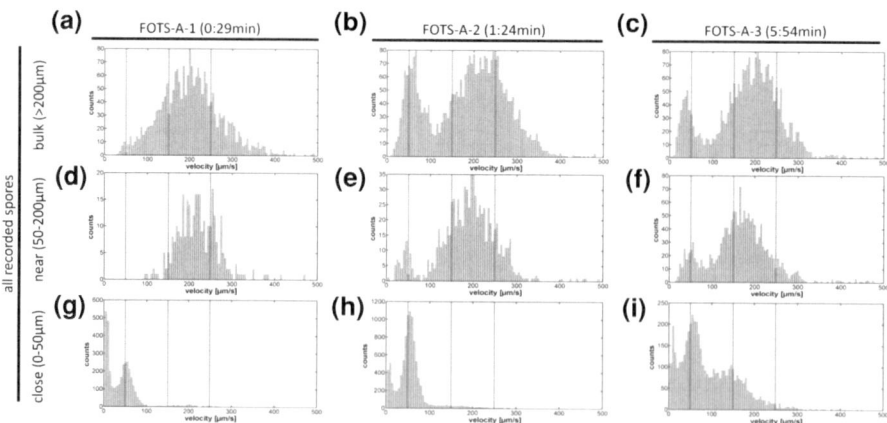

Fig. A.47 Velocity histograms for on FOTS-A-* shown for three distances from the surface **a–c** 200–800 µm, **d–f** 50–200 µm and **g–i** 0–50 µm. **a, d, g** is analyzed after 0:29 min, **b, e, h** after 1:24 min, and **c, f, i** after 5:54 min. The data shown in **a–c** is already discussed in Sect. 5.1.1 under the name Bulk-III-A-*. The *verticals lines* indicate velocities of 50, 150 and 250 µm/s

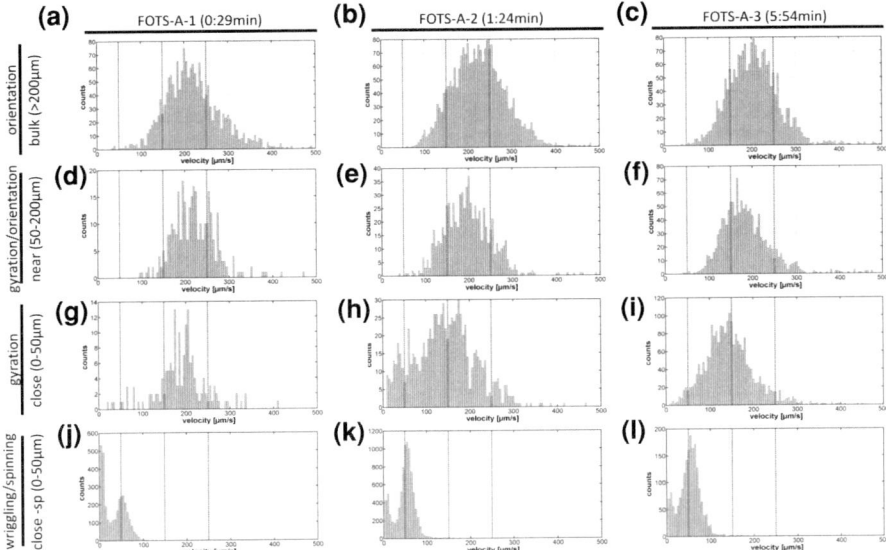

Fig. A.48 Velocity histograms for the spores assigned to *orientation* pattern **a–c**, to the spores assigned to the *gyration* pattern **d–i** and to the spinning pattern **j–l** on FOTS-A-*. The velocity histograms are shown for three distances from the surface **a–c** 200–800 μm, **d–f** 50–200 μm and **g–l** 0–50 μm. **a, d, g, j** are analyzed after 0:29 min, **b, e, h, k** after 1:24 min and **c, f, i, l** after 5:54 min. The data shown in **a–c** is already discussed in Sect. 5.1.1 under the name Bulk-III-A-*. The *verticals lines* indicate velocities of 50, 150 and 250 μm/s

due to surface interactions of the spores with the FOTS surface. The peak around 5 μm/s can be assigned to spores *sticking* (phase II of the *hit and stick* pattern) on the surface. The peak around 50 μm/s is assigned to *spinning* spores on the surface. It is possible to exclude that the peak around 50 μm/s is caused by spores swimming according to the *wobbling* pattern [see comparison Fig. A.47, panel (g), all spores and Fig. A.48, panel (j), spores assigned to the *hit and stick* pattern]. The amount (small number of counts in the histogram) of spores assigned to the *gyration* pattern at this time (FOTS-A-1) is extremely small [see Fig. A.47, panel (g) and Fig. A.48, panel (g)].

With elapsing time the velocity distribution for spores belonging to the *hit and stick* and *spinning* pattern changes *close* to the surface [Fig. A.48, panels (j, k, l)]. After 1:24 min [panel (k), FOTS-A-2] the ratio of *sticking* and *spinning* spores is inverted compared to the situation after 0:30 min (FOTS-A-1). The situation after 5:54 min [panel (l), FOTS-A-3] is similar to the situation in FOTS-A-2. Both peaks which are assigned to *sticking* spores and *spinning* spores are observed. Based on the knowledge of the individual trace analysis (Sect. A.3.1) in FOTS-A-3 [panel (l)] the sticking spores are not caused by spores assigned to the *hit and stick* pattern (which does not occur anymore during FOTS-A-3) but rather caused by the one settling spore (see Fig. A.45).

The velocity distribution for the spores assigned to the *gyration* pattern also changes significantly with elapsing time. Shortly after the injection nearly no spores assigned to the *gyration* pattern are detected *close* to the surface [Fig. A.48, panels (g–i)]. This is best seen comparing the velocity histogram of all analyzed spores [Fig. A.47, panels (g–i)] with the velocity histogram showing only the spores assigned to the *gyration* and *orientation* pattern [Fig. A.48, panels (g–i)]. Furthermore, in the latter histograms it is shown that with increasing observation time the amount of spores assigned to the *gyration* pattern strongly increases. At the last analyzed point in time [see Fig. A.47, panel (i), FOTS-A-3] the spores assigned to the *gyration* pattern are clearly observed as an individual peak in the velocity histogram of all spores. The occurrence of these spores clearly marks a change in the exploration behavior on FOTS.

For the slow spore fraction the velocity distribution *near* (50-200 µm) the surfaces [Fig. A.47, panels (d, e, f)] also changes with increasing time. Shortly after the injection no slow spores are detected [panel (d)]. At the intermediate time [panel (e)] a defined peak around 40 µm/s is visible. This peak is studied in detail to know whether it is caused by spores assigned to the *wobbling* pattern or by spores slowly leaving the surface. In Fig. A.49 the spore trajectories corresponding to the histogram [panel (e)] are shown. Two wobbling spores are detected in this figure. One spore swims towards the surface ("slow 1"). The other spore swims in a distance of 190 µm parallel to the surface ("slow 2"). Therefore the peak in the histogram is caused by *wobbling* spores and not by spores leaving slowly the surface.

The spores assigned to the *gyration* pattern are slower when they swim *close* (0–50 µm) to the surface [Fig. A.48, panels (h, i)] than when they swim *near* (0–200 µm) the surface [Fig. A.48, panels (e, f)]. This observation is not made shortly after the injection [panels (d, g)] which might be due to the low statistic at this time [only three spores are assigned to the pattern (see Table A.7, plus the approaches/detachments of spores assigned to the *hit and stick* pattern)]. With elapsing time and the increasing amount of spores assigned to the *gyration* pattern the difference between the swimming speed *close* and *near* the surface increases.

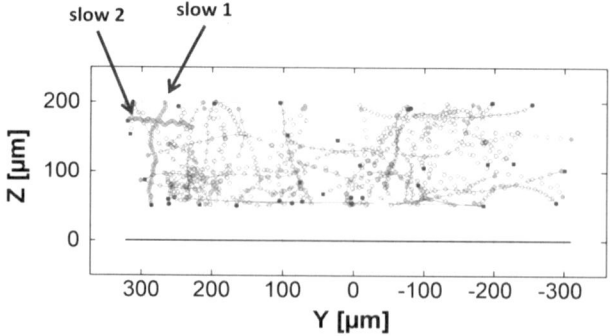

Fig. A.49 Spore trajectories *near* the surface (50–200 µm) color coded depending on different velocities *black*, filled marker: start and end point of the trajectory; *gray*, filled marker: v < 50 µm/s; *gray*, unfilled marker: v > 50 µm/s

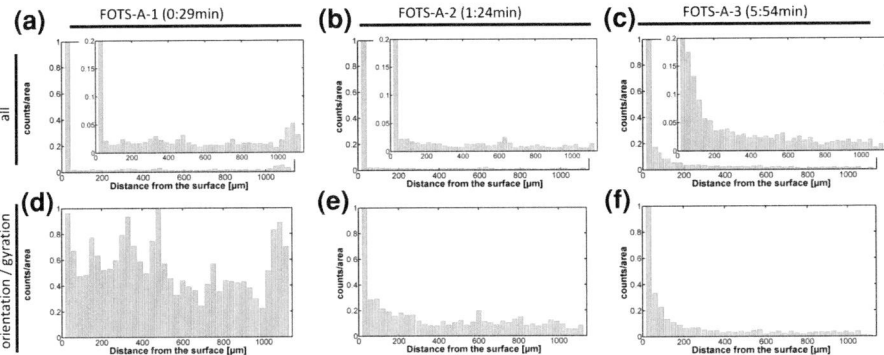

Fig. A.50 Distribution in the observation volume on FOTS-A-*. **a–c** shows all recorded spores with two different y scales and **d–f** only the spores assigned to the *gyration* and *orientation* pattern

In Fig. A.50 the distribution of the spores is shown for the whole observation volume. All analyzed spores are shown in Fig. A.50, panels (a–c) whereas in Fig. A.50, panels (d–f) only the spores assigned to the fast spore fraction in the bulk and to the *gyration* pattern are shown. The distribution of all spores is strongly influenced by the *hit and stick* and *spinning* pattern staying a long time on the surface. To show the distribution in solution the identical distribution is plot with a different scale on the y-axis. For the last experiment [panel (c)] the distribution changes and a spore accumulation is detected within the first 200 μm. To investigate the exploration behavior the distribution of the spores assigned to the *gyration* and *orientation* pattern is more important than the spores belonging to the *wobbling* pattern because these are the spores which actively search for a place to settle. The spores assigned to the *spinning* pattern have already selected a position on the surface where they might be settling and therefore these trajectories do not provide information how the spore achieved to get to this position. For this analysis the approach and detachment part of each spore trajectory belonging to the *hit and stick* pattern is assigned to the *gyration* pattern. For this spore fraction [Fig. A.50, panels (d–f)] the distribution changes significantly with elapsing time. In the beginning no accumulation of spores is detected in the vicinity of the surface. This can be explained by the fact that all spores which get close to the surface are trapped at the interface. With elapsing time the spores do not have to stick to the surface and are able to explore the surface. At the last experiment [FOTS-A-3) the typical [as observed for AWG (Sect. A.1.2) and PEG (Sect. A.2.2)] spore accumulation is observed in the vicinity of the surface. This enrichment is detected at 220 μm [see Fig. A.50, panels (f)]. In Fig. A.50, panel (f) the spore distribution is very similar to the distribution on AWG (see Sect. A.1.2, and Sect. 6.3 for discussion of the comparison between AWG and FOTS).

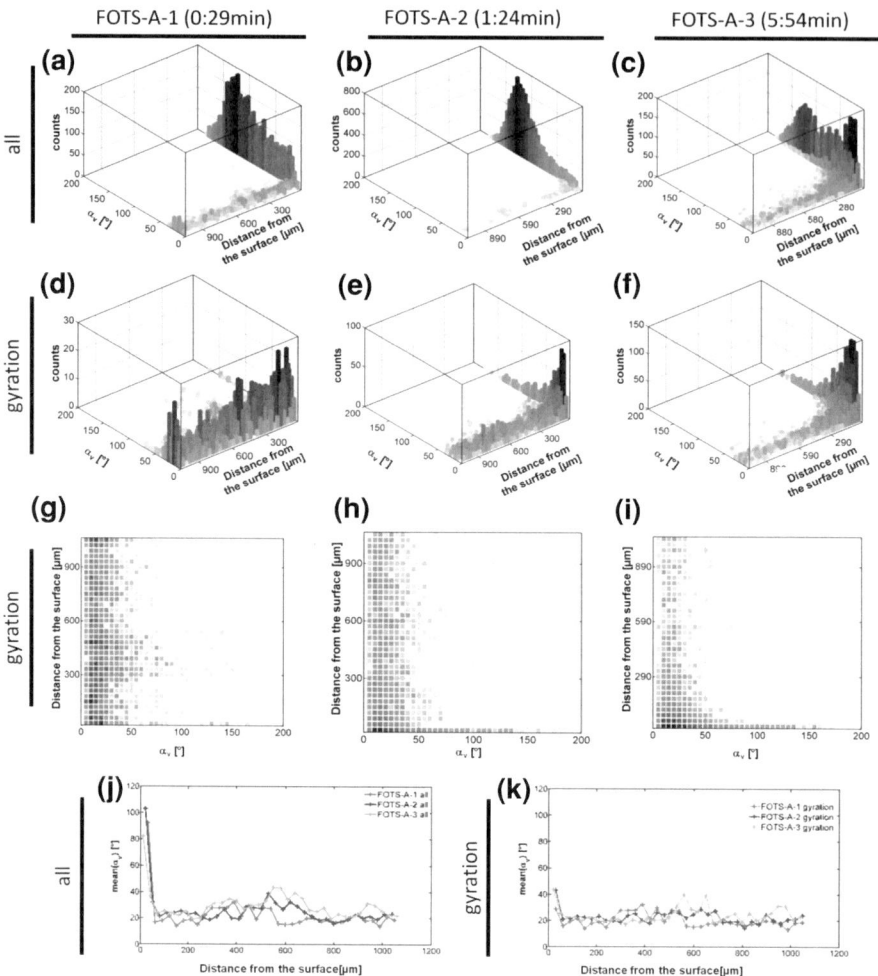

Fig. A.51 The α_v distribution for the exploration on FOTS-A-*. **a–c** all analyzed spores; **d–i** spores assigned to the *gyration* pattern; **j** $\bar{\alpha}_v$ for all spores; **k** $\bar{\alpha}_v$ for the spores assigned to the *gyration* and *orientation* pattern; **j, k** *gray curve*: FOTS-A-1, *black curve*: FOTS-A-2, and *light gray curve*: FOTS-A-3; **a, d, g** FOTS-A-1 (0:29 min); **b, e, h** FOTS-A-2 (1:24 min); **c, f, i** FOTS-A-3 (5:54 min); **a–f** 3D histogram of α_v; **g–i** 2D top view of the perspective 3D histogram. The count rate is encoded in the height of the bars, the darker the bar the higher the count rate

Since the spore distribution changes with elapsing time the experiments FOTS-A-1 to FOTS-A-3 are analyzed separately for the following detailed exploration analysis. Different parameters (Fig. A.51: α_v, Fig. A.52: α_z and Fig. A.53: velocity) are shown. Panels (a, d, g) shows experiment FOTS-A-1, panels (b, e, h) shows experiment FOTS-A-2 and panels (c, f, i) shows experiment FOTS-A-3. Figures A.51, A.52 and A.53, panel (j) shows the mean value of the corresponding

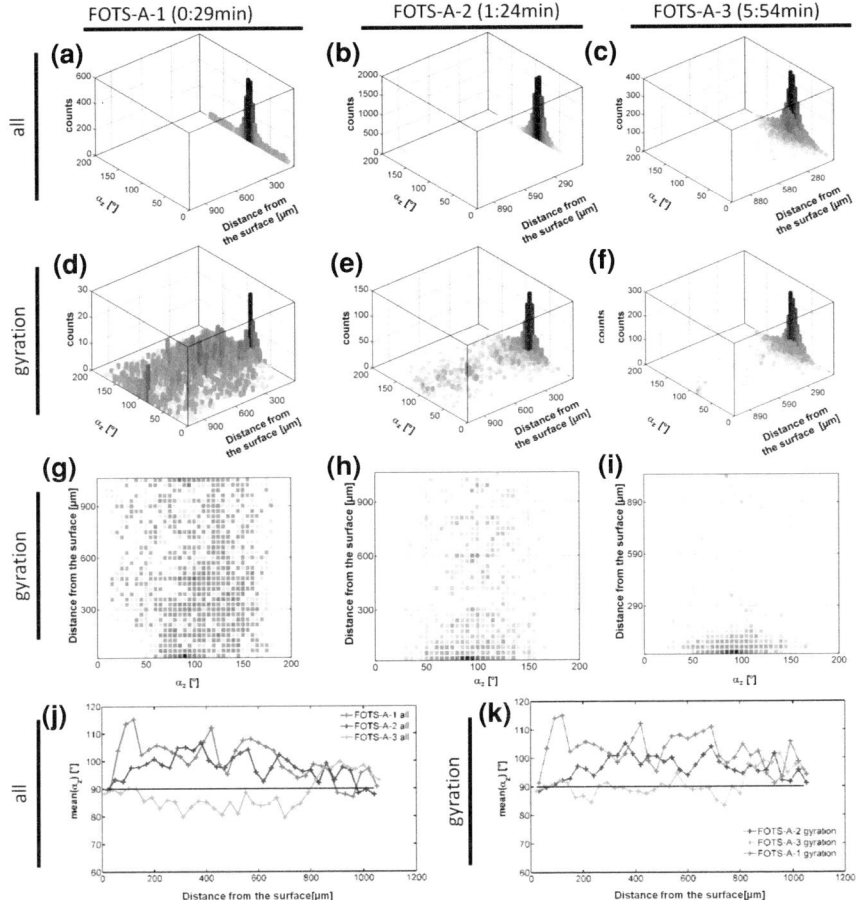

Fig. A.52 The α_z distribution for the exploration on FOTS-A-*. **a–c** all analyzed spores; **d–i** spores assigned to the *gyration* pattern; **j** α_z for all spores; **k** $\bar{\alpha}_z$ for the spores assigned to the *gyration* and *orientation* pattern; **j, k** *gray curve*: FOTS-A-1, *black curve*: FOTS-A-2, and *light gray curve*: FOTS-A-3; **a, d, g** FOTS-A-1 (0:29 min); **b, e, h** FOTS-A-2 (1:24 min); **c, f, i** FOTS-A-3 (5:54 min); **a–f** 3D histogram of α_z; **g–i** 2D top view of the perspective 3D histogram. The count rate is encoded in the height and in the color of the bars, the darker the bar the higher the count rate

Figure. for all analyzed spores and panel (k) shows the mean value for the spores assigned to the *gyration* pattern and to the fast spore fraction in the bulk. In panels (j, k) the red curve represents the spores from experiment FOTS-A-1, the blue curve from experiment FOTS-A-2 and the green curve from the experiment FOTS-A-3. In Figs. A.51, A.52 and A.53, panels (a–c) all analyzed spores are shown whereas in panels (d–i) only the spores assigned to the *gyration* pattern and to the fast spore fraction are shown.

In Fig. A.51 the α_v distribution is shown. For all analyzed spores [panels (a–c)] the spores on the surface (*spinning* or *sticking*) dominate the distribution. For the

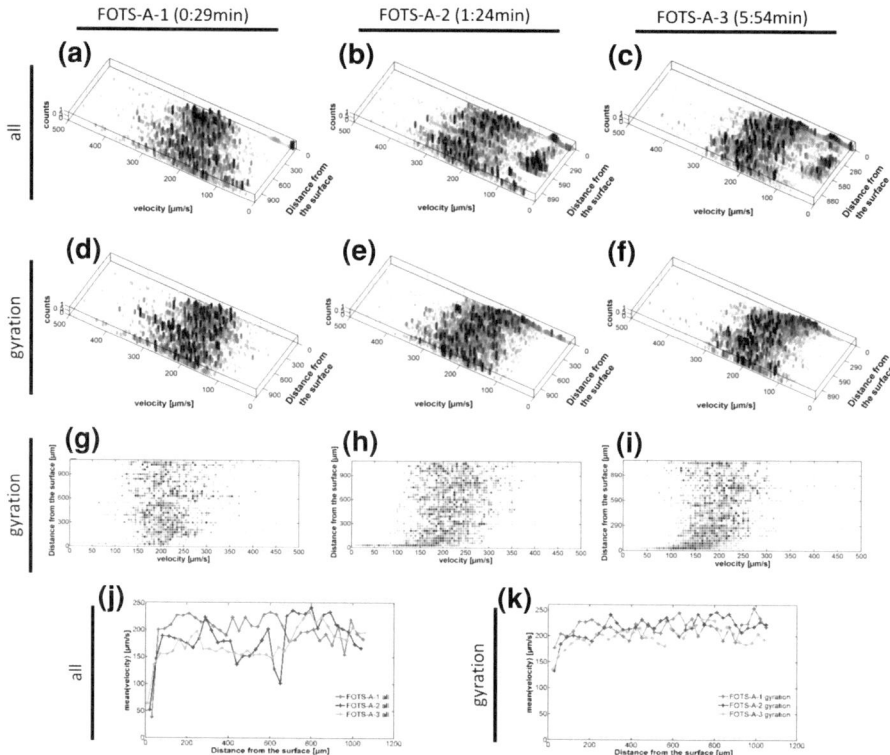

Fig. A.53 Velocity distribution for the exploration on FOTS-A-*. **a–c** all analyzed spores; **d–i** spores assigned to the *gyration* pattern; **j** $\bar{v}$ for all spores; **k** $\bar{v}$ for the spores assigned to the *gyration* and *orientation* pattern; **a, d, g** FOTS-A-1 (0:29 min); **b, e, h** FOTS-A-2 (1:24 min); **c, f, i** FOTS-A-3 (5:54 min); **a–f** 3D histogram **g–i** 2D top view of the perspective 3D histogram. The count rate is encoded in the height of the bars, the darker the bar the higher the count rate

spores assigned to the *gyration* and *orientation* pattern [panels (d–i)] the distribution changes with elapsing time. This is best seen in the top view [panels (g–i)]. For experiment FOTS-A-1 [panel (g)] no change in respect to the surface distance is detected for the α_v distribution. The distribution *close* to the surface is the same as the one for the bulk. This means that the spores do not perform more turns when they get closer to the surface. The α_v distribution shown in Fig. A.51, panel (g) even gives the impression that the angular distribution is narrower between 50–250 μm from the surface. This would mean that the spores in that segment prefer to swim straight towards the surface.

With elapsing time and with the occurrence of the *gyration* pattern the α_v distribution changes [see Fig. A.51, panels (g, i)]. For the experiment FOTS-A-3 [panel (i)] the α_v distribution increase in a distance of 200 μm from the surface. These observations are also made for the $\bar{\alpha}_v$ distribution [panel (k)] although not as pronounced as for the α_v distribution shown in panels (g, i).

Figure A.52 shows the α_z distribution. The α_z distribution changes significantly with elapsing time. This is best seen for the $\bar{\alpha}_z$ distribution in Fig. A.52, panels (j, k). For the experiment FOTS-A-1 a strong preference for all spores is to swim towards the surface ($\alpha_z > 90°$). This preference is lost with elapsing time (FOTS-A-2: blue curve and FOTS-A-3 green curve). For experiment FOTS-A-3 (green curve) there is a general tendency detected that all analyzed spores [panel (j)] swim away from the surface whereas for the spores assigned to the *gyration* and *orientation* the same number of spores swim towards and away from the surface. Only around a distance of 200 μm to the surface more spores swim away from than towards the surface.

Figure A.53 shows the velocity distribution for the spore exploration on FOTS. For the spores assigned to the *gyration* and *orientation* pattern (d–i, k) the distribution also changes with elapsing time. For FOTS-A-1 [panels (g, k (red curve))] the velocity is only slightly slower when the spores approach the surface than compared to the observed velocity in the water column. For the last experiment [FOTS-A-3, panels (i, k (green curve))] a slowdown of the swimming speed is detected 200 μm from the surface. This effect is not seen for all spores because the slow spore fraction disturbs the effect.

In Fig. A.54 the $\bar{\alpha}_z$ distribution is shown in detail for each experiment [FOTS-A-1 panel (a), FOTS-A-2 panel (b) and FOTS-A-3 panel (c)] and for the spores assigned to the *gyration* and orientation pattern. The $\bar{\alpha}_z$ distribution in each panel is plotted for all analyzed spores (blue), the spores swimming towards the surface (red) and the spores swimming away (green). Figure A.54 shows that the spores analyzed in FOTS-A-1 [panel (a)] a preference to swim towards the surface is observed for the whole observation volume (blue curve).This flow exist only during the analysis of FOTS-A-1. With elapsing time and the disappearance of the *hit and stick* pattern the flow towards the surface also vanishes. A comparable flow towards the surface is neither observed for AWG nor PEG.

Furthermore, while the analysis of FOTS-A-1 Fig. A.54, panel (a) in vicinity (0–100 μm) the spores swim steeper towards the surface (red curve) than the spores that swim away (green curve). Only shortly after the injection [FOTS-A-1, panel (a)] a peak in the α_z distribution at a distance of 90 μm from the surface is detected. This peak means that the spores swim directly towards the surface. With elapsing time the spores approach the surface [red curves, panels (b, c)] less steep and the distribution for the spores swimming away from the surface gets steeper. For FOTS-A-3 [panel (c)] no difference is detected between the steepness of the approach (red curve) and of the detachment (green curve) from the surface.

In Fig. A.55 the velocity distribution is studied in detail for the spores belonging to the *gyration* and *orientation* pattern. Experiment FOTS-A-1 is shown in panel (a), FOTS-A-2 in panel (b) and FOTS-A-3 in panel (c). The velocity distribution is divided into three sections: all analyzed spores (blue), the fraction which swims towards the surface (red) and the fraction which swims away from the surface (green). In Fig. A.55 the spores in the bulk swim as fast towards the surface as away from the surface. But close to the surface the situation is different. For FOTS-A-1 [panel (a)] the spores swim faster away from the surface than the

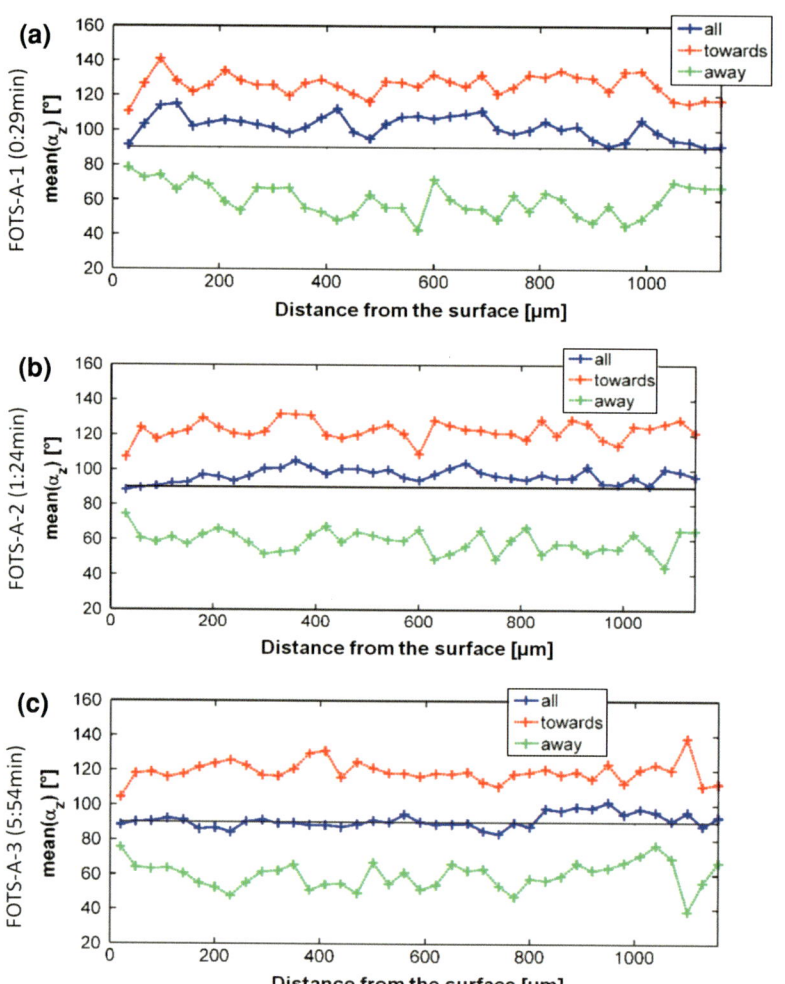

Fig. A.54 Detail analysis of $\bar{\alpha}_z$ for the spores assigned to the *gyration* and *orientation* pattern. FOTS-A-1 **a**, FOTS-A-2 **b** and FOTS-A-3 **c**. The *blue curve* represents all spores, the red curve the spore fraction which swims towards the surface and the *green curve* the spore fraction which swims away from the surface

spores swim towards the surface. Indeed there is a minimum detected for the approach. This minimum coincides with the peak in the α_z distribution for the spores swimming towards the surface [see Fig. A.54, panel (a)]. This drop in velocity can be explained by spores swimming a turn, precisely, the spores perform a turn towards the surface.

With increasing experiment time [panels (b, c)] the spores swimming away from the surface are slower than the spores swimming towards the surface. In Fig. A.55, panel (b) this effect is observed up to a distance of 200 μm from the

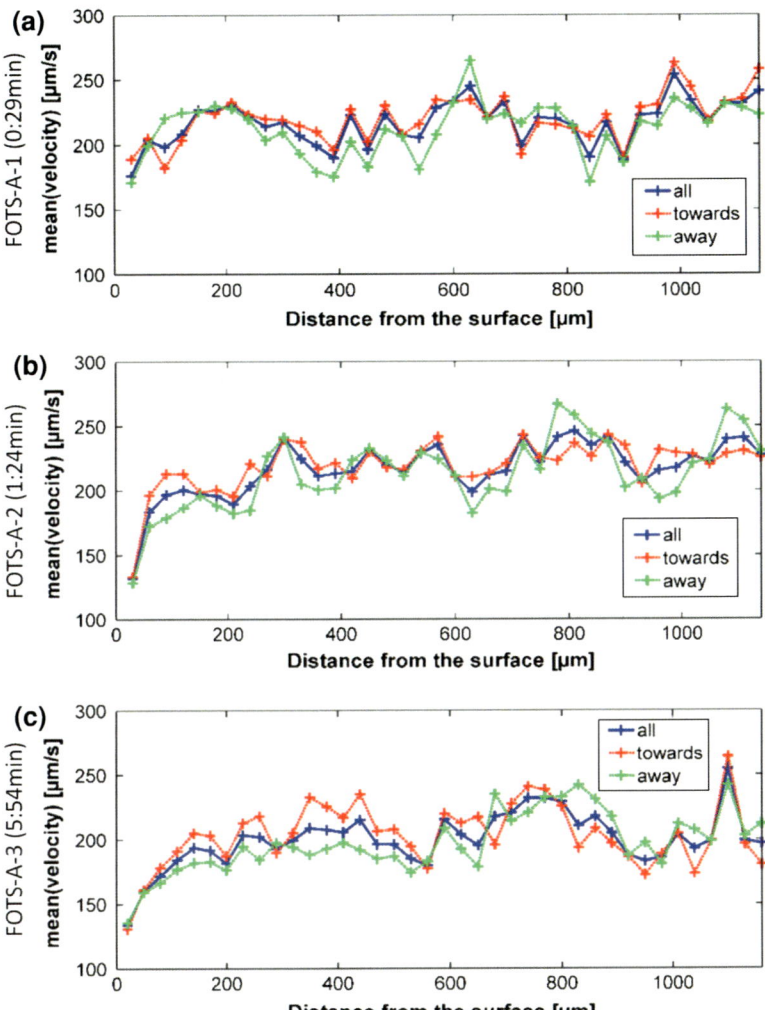

Fig. A.55 Detail analysis of $\bar{v}_v$ for the spores assigned to the *gyration* and *orientation* pattern. FOTS-A-1 **a**, FOTS-A-2 **b** and FOTS-A-3 **c**. The *blue curve* represents all spores, the *red curve* the spore fraction which swims towards the surface and the *green curve* the spore fraction which swims away from the surface

surface, whereas in Fig. A.55, panel (c) it can be seen up to a distance of 580 µm from the surface.

In summary, the exploration behavior in FOTS is complex and changes with elapsing time. In the beginning no surface exploration is observed and all spores which come close to the surface are trapped on the surface. With elapsing time this behavior changes and the surface is explored similar to an AWG surface.

References

1. M. Heydt, A. Rosenhahn, M. Grunze, M. Pettitt, M.E. Callow, J.A. Callow, J. Adhes. **83**(5), 417–430 (2007)
2. M.E. Callow, J.A. Callow, J.D. Pickett-Heaps, R. Wetherbee, J. Phycol. **33**(6), 938–947 (1997)

Index